Praise for *Challenged by Carbon*:

"The author's enthusiasm leaps out of every page ... readable, jargon–free and informative book on climate change ... sets the present in the context of past changes. Anecdotes, personal reminiscences and clear science will captivate and inform the general reader and may well offer new insights to the specialist. A really good read."

   Lord Oxburgh, House of Lords Science and Technology Committee

"As a geologist Lovell gives an authoritative insider's view of the oil industry's various approaches to climate change and the contribution industry can make through carbon capture and storage. Enlivening the book with geological insights, he also maps out the government frameworks needed to meet the climate challenge."

   Sir Mark Moody-Stuart, former Chairman of Royal Dutch/Shell and Anglo American plc

"Bryan Lovell's voice is a new one on the climate change stage. Compelling, lucid, and enjoyable – this book demystifies geology for the non-specialist and elucidates how geologists and the oil industry can contribute solutions to the problem of global climate change."

   Professor Robert Socolow, Princeton University

"Bryan Lovell has produced a remarkable book which draws on a lifetime of experience in the oil industry to identify new and creative ways of dealing with the challenge of global warming. This is a book which deserves the widest audience not just within the scientific community and the energy industry but also at the highest levels of policy making."

   Nick Butler, University of Cambridge and Senior Policy Adviser, 10 Downing Street

"Bryan weaves a compellingly entertaining story – the Oil Industry's change in attitude to carbon is well documented, as is the capability for 'putting the carbon back'. The book then highlights the author's frustration at the continued absence of an international regulatory regime that is capable of addressing the real objective function."

   Dr David Jenkins, Director of BHP Billiton plc and former Director Technology BP plc

"The central theme of *Challenged by Carbon*, that the oil and gas industry is a vital part of the solution as we transition slowly to a lower carbon energy future, is one that I heartily endorse."

   Professor Scott W. Tinker, University of Texas at Austin, Stage Geologist of Texas and former President of the American Association of Petroleum Geologists

Challenged by Carbon: The Oil Industry and Climate Change

Is there a low-carbon future for the oil industry?

Faced with compelling new evidence from the rocks that lie at the heart of its business, the petroleum industry is no longer able to ignore the consequences of climate change brought on by consumption of its products. Yet across the globe we will continue to need to burn fossil fuels as we manage the transition to a low-carbon economy.

As a geologist, oil man, academic and erstwhile politician, Bryan Lovell is uniquely well placed to describe the tensions accompanying the gradual greening of the petroleum industry over the last decade. He describes how, given the right lead from government, the oil industry could play a crucial role in stabilising emissions through the capture and underground storage of large volumes of carbon dioxide.

Lovell challenges entrenched prejudice on both sides of the debate between the environmentalists and the oil industry, giving a glimpse of oil barons as prospective environmental saviours rather than traditional stage villains. Ultimately he assigns major responsibility to us as consumers and to our elected governments, highlighting the need for decisive leadership and urgent action to establish an international framework of policy and regulation.

BRYAN LOVELL holds B.A. and M.Sc. degrees in geology from the University of Oxford and a Ph.D. from Harvard University. Following 12 years as a lecturer in geology at the University of Edinburgh and as a consultant to the oil industry, he worked for BP Exploration from 1981 to 1996, joining as Chief Sedimentologist, and subsequently holding positions as Exploration Manager and General Manager Ireland, International Exploration Manager with special responsibility for Middle East, and Head of Recruitment, BP Group. He is currently a Senior Research Fellow in Earth Sciences at the University of Cambridge, working on controls exercised by mantle convection on the elevation of Earth's surface, and continues to provide consultancy advice to the oil industry. Dr Lovell was the Scottish Liberal Party energy spokesman from 1978 to 1979 and ran as a parliamentary candidate in 1979, finishing third out of five behind Michael Ancram and Gordon Brown. He was awarded an OBE in 1989 for services to Anglo-Irish relations and has recently been elected President-designate of The Geological Society of London (2010–2012).

Challenged by Carbon

The Oil Industry and Climate Change

BRYAN LOVELL

CAMBRIDGE UNIVERSITY PRESS

Cambridge, New York, Melbourne, Madrid, Cape Town, Singapore,
São Paulo, Delhi

Cambridge University Press
The Edinburgh Building, Cambridge CB2 8RU, UK

Published in the United States of America by
Cambridge University Press, New York

www.cambridge.org
Information on this title: www.cambridge.org/9780521197014

First published 2010

Printed in the United Kingdom at the University Press, Cambridge

A catalogue record for this publication is available from the British Library

ISBN 978-0-521-19701-4 Hardback
ISBN 978-0-521-14559-6 Paperback

Contents

Acknowledgements

Young geologists learn by looking at rocks in the company of older geologists. Older geologists keep learning about rocks by looking at them in the company of young geologists. All geologists learn by looking at rocks with people who aren't even sure what geology is. So over the years I have become indebted to many people, several of them characters in the story told here: thanks to all of you.

Then there are those who have helped quite specifically with the preparation of this book, through some combination of encouragement, advice, gift of materials and warnings. My thanks go to: Francisco Ascui, Claire Budd, Nick Butler, Andy Chadwick, Anthony Cohen, Harry Elderfield, Susan Francis, Gardiner Hill, James Jackson, David Jenkins, Jeremy Leggett, Heather Lovell, Dan McKenzie, Mark Moody-Stuart, Ted Nield, Richard Norris, Heather Poore, Robert Socolow, Bob White, Nicky White and Eric Wolff.

We geologists prefer to spend most of our time outside, looking at rocks in many places and in all weathers. When measuring sandstones on an exposed cliff and recording their special features in a field notebook, it is a big help to have a resolute companion with neat handwriting. It is even better if she can see beyond the cliff. This book is for Carol.

Preface

Within the last decade the scientific case for anthropogenic climate change has become significantly more plausible. New observational science offers crucial support for the computer-based speculations of those creating models of climate change. The record of Earth's past climates recorded in rocks and ice can now be measured with far greater definition than before: divided into thousands rather than millions of years. This major break-through means that changes in climate that took place long ago can now reasonably be compared with those seen in the recent past.

One of these past changes in climate is a particularly important guide to present-day concerns: a dramatic warming event that took place 55 million years ago (55 Ma). Comparison of the volume of carbon released to the atmosphere 55 million years ago and the volume we are now releasing ourselves strongly suggests that we are indeed facing a major global challenge. We are in danger of repeating that 55 million-year-old global warming event, which disrupted Earth for over 100 000 years. That event took place long before *Homo sapiens* was around to light so much as a campfire. Now we have no excuses, we are here and we are aware of our capacity to precipitate major inimical changes to our habitat on this planet. We can cope, but only by adopting a new intellectual framework for energy policy that is based on that awareness.

This is an unusual challenge to the established order, comparable to the greatest periods of political and social change.

Successful resolution will require an unusual degree of cooperation between all sorts of tribes: academic, social, financial, industrial, political and national. This kind of cooperation was the real value of the 1997 Kyoto Protocol and is the hope for the crucial successor meeting in Copenhagen in 2009. The Kyoto agreement was never going to be a sufficient answer in itself to coping with climate change, but it was a sign that the global community has the capacity to edge towards the scale of cooperation that is required.

That cooperation clearly has to embrace China and India. These two countries are moving along paths of development that emulate those followed previously in the developed world, with heavy reliance on fossil fuels – especially coal. If the source of energy for development in China and India remains as it is now, the per-capita emission of carbon in those countries will continue to increase rapidly. On a planet with a forecast increase in human population measured in billions during this century, per-capita measurements become ever more significant. At the beginning of the twenty-first century the per-capita emissions in China and India were an order of magnitude lower than those in North America, Europe and Russia: now that gap is closing. Can the developing countries be helped to achieve their aspirations for rapid development while maintaining a low per-capita output of fossil carbon? Can the developed countries maintain the confidence of their consumers and voters while reducing per-capita output of fossil carbon?

Here the oil companies may have a chance of redemption from their classical role as the villains of climate change, by giving a positive response to being challenged by carbon. In principle they could capture and then store safely underground a good part of the fossil carbon released to the atmosphere through their agency – and that of the coal industry – although the price in energy and dollars of that capture and safe storage is still not clearly defined.

Emissions of greenhouse gases by the oil companies may be divided into those resulting from their own operations (a little over 10 per cent of the total), and the remaining quantity (approaching 90 per cent of the total) that is released by the use of their products by their customers. The oil companies have made considerable efforts in recent years to control greenhouse gas emissions from their own operations, not necessarily to their commercial disadvantage. Responsibility for coping with the far greater quantity of emissions resulting from the use of oil company products by customers has yet to be assigned.

At this stage, a series of broad questions may be framed, by no means complete, but an indication of the scale of change required:

Can the major international non-state oil companies, who control only a few per cent of the world's reserves of oil and gas, persuade their shareholders to keep investing when they seek to make money by disposing of fossil carbon (in the form of anthropogenic carbon dioxide), as well as profit by pumping it out of the ground (in the form of oil and natural gas)?

Can the major state oil companies, who control the greater part of the world's reserves of oil and gas, persuade their governments that part of their role should be the safe disposal of carbon dioxide?

Can political, economic and financial institutions adapt to a global imperative to regard the safe capture and disposal of carbon dioxide as an activity as important as taking fossil fuels out of the ground?

Can cutting demand for energy be given similar emphasis to ensuring supply?

Can a representative democracy accept as an overriding basis for policy an intellectual framework that demands a perspective of fifty years rather than five?

Can both governments and governed come to regard activities that restrain per-capita output of carbon as a social good, on a par with provision of health services and education?

It is claimed in this book that all this could be achieved, given a widespread deep conviction that there really is a problem to be solved. We already have to hand the technology to give ourselves breathing space. So the six questions posed above hinge on yet another question: how can that depth of conviction be achieved? It is argued here that the study of rocks can provide, from observational science rather than from computer-based models, the necessary depth of intellectual conviction that there is a problem in the first place. Then the significant role that subsurface storage of carbon dioxide could play in resolving that problem is assessed, with particular reference to the future of the oil industry.

This book is written from the individual perspective of a geologist with a range of experience in universities, industry and politics. Specifically it is based on a working knowledge of the rocks in which oil is found and in which carbon dioxide could be stored. Broadly it is based on the expertise of those I have been lucky enough to have as teachers, colleagues and students. This large debt is acknowledged throughout the narrative. In that respect, readers familiar with operations in the oil industry might now anticipate a homily on team behaviour. They can relax a little: the need for intense cooperation between different tribes to cope with climate change should be obvious enough from the story itself.

Lying between mainland Europe and the UK is the source of much late-twentieth-century prosperity: the North Sea. The story told in this book has this oil province and those working there as a recurrent theme. It was among those with experience in the North Sea that much of the crucial change in thinking took place. For environmentalists and oil folk alike it is an important source of evidence and ideas. The many Earth scientists who are involved in energy and the environment have a common meeting place, Burlington House in Piccadilly, London. There the Geological Society of London belies the age of its building, and its status as the world's oldest geological organisation, by mixing academics and industrialists in politically alert discussion and debate on topics of contemporary public significance. Reports of innovative activity on climate change are published by the Society for the benefit of its Fellows: these reports are gratefully drawn on here for presentation to a wider audience.

There has been an apparent contrast with the story in North America, a distinction now happily fading. Until recently an environmentally alert European visitor to Alberta and Texas would have found disappointingly little concern within the oil community that their industry might be a significant part of a serious and developing problem with climate change. Fossil carbon was taken from the ground with increasing enthusiasm from ancient wells in the Permian Basin of Texas at times of rising prices and was cheerfully passed to the customer for eventual release to the atmosphere as carbon dioxide. Canadian tar sands were mined using a deal of water and energy to release the sticky oil, with thoughts of profit apparently running well ahead of environmental concerns. These attitudes have until recently been reflected in the leadership and policy of professional organisations such as the American Association of Petroleum Geologists (AAPG). Those of us involved with both the AAPG and the Petroleum Group of the Geological Society in London have had plenty of contrasting literature to consider and debate over the last decade.

The key issues of climate change will eventually be resolved in one way or another on a global scale, not just in Europe and North America. Fortunately a natural virtue of the oil industry is that it is obliged to be international. Oil and gas are not distributed evenly on this planet and most of the remaining reserves lie outside the North American and European homelands of the major non-state oil companies. Only part of Earth was covered by the ancient Tethys Oceans, the geological evolution of which led to the concentration of oil in what is now the Middle East. Will the oil industry be able to seize the advantage of its global perspective to bring general environmental benefit to its customers, while protecting its own profits? We have long relied on the oil folk to use their ingenuity to supply us with their mighty handy products: now we need their inventiveness to help us manage our transition from that dependency.

Is the industry prepared to help? The scale of the problem is certainly recognised. The view of Shell was given by their Vice President CO_2, Mr Bill Spence, in a public lecture at the Geological Society in London on 4 November 2008 (Spence, 2008). By the middle of this century we shall need to supply a much larger global population, aspiring to greater wealth, with double the amount of energy used today. All that energy will have to be supplied while emissions of carbon dioxide are severely restricted: the huge numbers involved are the subject of much of this book. The tribes will certainly need to combine to tackle this extraordinary task. Not least of those tribes is the oil industry.

— Before we take these issues further, one important matter needs to be faced. Is the earnest enquiry set out in this book entirely misplaced from the very start? Are we already too late? Should we be concentrating on how to cope with the apocalypse, because we have little hope of avoiding it?

There are those, notably that doyen of the environmentalists, Dr James Lovelock (2006, 2009), who consider that we have abused the planet beyond hope of redemption. In this gloomy view of our prospects, all we can do is prepare to act defensively as climates change dramatically, sea-level rises, and mass migrations and the collapse of societies test our species to the limit. Possible defensive actions taken in these dire circumstances would differ in significant respects from the actions humankind might take to prevent the very occurrence of such disasters: seeking a balance between mitigation and adaptation (Bierbaum, Holdren, MacCracken, Moss and Raven, 2007; Hunt, 2009) would no longer be a priority.

Whatever the merits of considering a range of defensive actions at this stage, one would have to be very sure that disaster was inevitable before ceasing to try to avoid it. A premise of this book, based on a reading of the reality written in the rocks, is that we cannot be confident that we have gone beyond Lovelock's point of no return. If across the globe we now unite in wise action, we can still turn back from the edge; if not to stability, at least away from the apocalypse.

This book attempts to explain in plain language how this relative optimism arises from the study of rocks. We geologists, like doctors and lawyers, seek to protect our income by using obscure language to describe quite ordinary things. I hope that such jargon is not obvious in this book, or that at least it is explained where it does creep in. For those who wish to study the scientific papers to which I refer, a word of warning. Jargon rules the day in some of these publications, although access to the largely comprehensible abstracts should in most cases be easy enough; I recommend simply typing the names of the authors into Google or Google Scholar, with the year of publication and a key word or two from the title.

A key point underpins this book: you can't argue with a rock. We can simply try to understand rocks by examining them carefully in the field and laboratory, using our wits and our imagination. Rocks are tangible objects that humans find useful for many purposes, including the provision of energy and the disposal of waste. So there is reality in rocks that we should strive to grasp.

In that spirit, Chapter 1 describes how a few refreshingly jargon-free geologists had a key influence on the relationship between the oil industry and the environmental movement at the crucial time of the Kyoto climate summit in 1997. Following that bit of modern history, we go way back in time. Chapter 2 shows how 55-million-year-old rocks record the reality of a warming event that is a salutary guide to our present concerns on climate change. This account draws on the emerging detailed understanding of that 55 Ma warming event on a human timescale. That 55 Ma perspective was available to participants in a 2003 international scientific conference that is considered in the last of the three introductory historical chapters; Chapter 3 has at its core the illuminating transcript of the March 2003 Geological Society debate in London on *Coping with Climate Change*, featuring two of the several Vice Presidents of BP and ExxonMobil. From a strict environmentalist's perspective this could be seen as a discussion between the damned and the devil, but here it is interpreted as the beginning of an important convergence across an earlier Atlantic Divide in the oil industry – a convergence that continues apace.

With the historical scene now set, Chapter 4 examines the strategic options open to the oil industry in reacting to today's growing scientific and political interest in climate change. The restraint imposed by rocks on all parties is identified. But do we have an intellectual framework within which this recent carbon challenge to the oil industry may be embraced in a new strategic outlook? Chapter 5 claims that we do have such a framework, and summarises a largely Princeton University-generated integration of engineering, economics and social science within which the oil industry can consider its strategic options. Carbon capture and storage (CCS) is picked out as a potentially significant contribution that the oil industry can make to controlling our release of carbon to the atmosphere, by putting back underground the carbon that the oil and coal industries have taken out for our eager use.

Can CCS be readily implemented by petroleum geologists and engineers? For them the principles and practice of CCS set out in Chapter 6 are second nature. Finally, how could CCS be made to happen? Chapter 7 places the onus firmly on government to set a framework of carbon policy and regulation, within which such vital activity as CCS can take place without beggaring everybody involved. This global regulatory framework for coping with the imminent carbon crisis should be well within the range of world leaders hardened by the financial crisis that began in 2008.

Chapter 8, 'The proof in the puddingstone', is a personal coda, connecting various events 55 million years ago to us. Puddingstone is an exceptionally hard rock, with a tough silica cement that probably formed as a result of the intense heat at Earth's surface during the 55 Ma warming event. The recent discovery of a hitherto elusive Roman puddingstone quarry north of London triggers a series of connections, including links to the 55 Ma reservoir sandstones at Forties field in the UK North Sea and to the carbon released to the atmosphere by our use of Forties oil.

For the Roman invaders of Britain settling in the Thames Valley a couple of thousand years ago, puddingstone was to become a key element in an essential technology: grinding corn. That particular imperial legacy now consists only of beehive querns and a few angular fragments of rock. This book says that our use of carbon cannot be allowed to become a millstone round the necks of our grandchildren – and it does not have to be. Our governments should give the putative environmental villains of the oil industry the chance to become carbon heroes: challenged by carbon yet not found wanting.

1

Geologists on the road to Kyoto

1.1 SCEPTICISM AND SCHOLARSHIP

Scholars are inclined to be sceptical, especially about the work of academic rivals. Awkward questions in seminars are admired, and established views are there to be challenged by new thoughts and new evidence. The academic cliché is that the passions run highest when the stakes are lowest. Some thirty years ago I had furious public arguments with distant colleagues, about the significance of variations in the thickness of sandstones formed hundreds of millions of years ago. These exchanges remained of very little interest to anybody but ourselves, even when later some of the ideas were put to use by the oil industry. But occasionally scientists do emerge blinking into the spotlight because they are arguing about something that really is important to everybody else. Around the turn of the century that became true for those studying climate change.

Just as climate change became a familiar headline in daily newspapers, as well as in academic journals such as *Science* and *Nature*, so the consensus amongst scientists that we did indeed have a problem became established (Oreskes, 2004). This is epitomised by the successive Reports of the Intergovernmental Panel on Climate Change (IPCC), which include the (1996) Report that immediately preceded the Kyoto Protocol of 1997. The IPCC is a scientific intergovernmental body set up by the World Meteorological Organisation and the United Nations Environment Programme: at time of writing its latest report was issued a couple of years ago (IPCC, 2007).

1

The notion of scientific agreement on climate change was spread to a wider public by former USA Vice President Al Gore in his film and book *An Inconvenient Truth* (2006). But once an issue such as climate change becomes public property, the habitual scepticism of scientists about conventional truths and consensus can be turned against them. This can be done with particular enthusiasm by those not habitually involved in the excruciating detail involved at the very core of the underlying scientific discussion. Climate-change sceptics such as Nigel Lawson (Lord Lawson) (2006, 2008) and Professor Bjorn Lomborg (2001), alerted by the implications for their own special areas of interest in economics, politics and finance, feel free to pick out the pips from fruity items of debate in the heavy crop of scientific agreement.

The public case for anthropogenic climate change, and a warming planet, has been made most strongly by climatologists using recent trends and computer forecasts. This focus on contemporary events has convinced the insiders, and the IPCC, but has dangers in carrying the case to the general public. Many people on this planet have livelihoods governed by weather. They are obliged to take a keen interest in weather day by day and season by season. They will therefore know at least something of the difference between weather and climate, and may with justice believe that they have already learned enough to take a view on present-day climate change. What happens to the scientists' case for anthropogenic climate change if it gets colder for a while, not warmer? It is conceivable – if perhaps unlikely – that the planet might be about to experience a natural, non-anthropogenic decade or so of global cooling that interrupts briefly the proven overall warming trend. Were that to happen, there would be public cries of triumph from the climate-change sceptics, and an even more elaborate global tour by Gore to explain that this was indeed just an interruption and, yes, we really did still have a problem.

At this point the scientific experts might reasonably become quite testy, sure in their belief that they are right and the public is being misled by unscrupulous characters who either don't understand the science properly or, even worse, are being deliberately misleading. To a degree that is happening already, even with the climatological data so strongly in favour of the case for

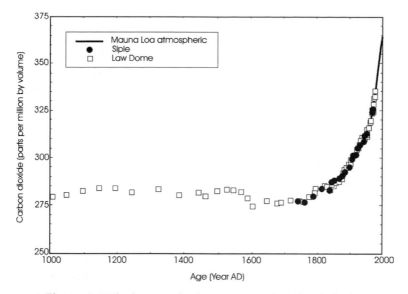

Figure 1.1 The increase in the concentration of carbon dioxide in Earth's atmosphere over the last 1000 years, in parts per million by volume. The historical evidence comes from gas trapped in Antarctic ice, supported by recent direct measurements of the atmosphere (see Figure 1.2). After a slide shown by Dr Eric Wolff of the British Antarctic Survey at the Geological Society meeting on *Coping with Climate Change* in March 2003.

anthropogenic climate change. To what evidence may the experts appeal to finally carry widespread conviction? A theme of this book, spelled out in the next chapter, is that the answer lies in the ground as well as in the skies.

We have some observational science to hand that no reasonable person can dispute. One observation is that the concentration of carbon dioxide in Earth's atmosphere has increased sharply since the Industrial Revolution (Figure 1.1, Figure 1.2) and that increase has been caused by us. Because carbon dioxide is a greenhouse gas, we would expect this increase in concentration to lead to higher temperatures at Earth's surface. The other observation is that the climate on this planet has changed a good deal in the past through 'natural' (non-anthropogenic) means, in some cases suddenly, long before we were around to understand any part of it (McManus, 2004) (Figure 1.3). Some of the more significant of these

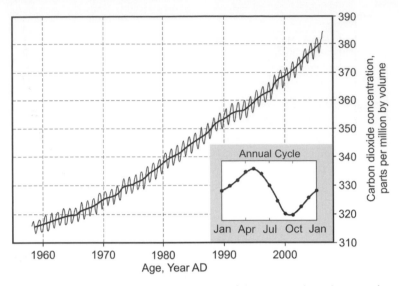

Figure 1.2 Detail from Figure 1.1 of the recent sharp increase in the atmospheric concentration of carbon dioxide recorded at Mauna Loa, Hawaii by Scripps Institution of Oceanography and US National Oceanic and Atmospheric Administration (NOAA) Earth System Research Laboratory. The annual cycle is not attributed to human activities, but we are held responsible for the overall trend.

changes appear to have been caused by 'natural' releases of carbon dioxide, a topic covered in some detail in Chapter 2.

There is a temptation to be fatalistic in reacting to this history of climate change in which we played no part. If all this change is going on anyway, why should we be concerned that we might now be interfering in some way? The answer comes from the geological record, and brings naught for our comfort. Study of past changes in climate, recorded in ice and rocks, has led many of us to believe that we are creating a serious potential problem for ourselves by releasing so much carbon dioxide into the atmosphere. A sudden warming event recorded 55 million years ago (55 Ma), caused by a natural, large and rapid release of carbon to the atmosphere, provides a particularly significant guide from the past to coping with our present concerns. The nature and origin of this 55 Ma event are discussed in Chapter 2. The official somewhat daunting official title of the event is Paleocene–Eocene Thermal Maximum, or PETM to its intimates.

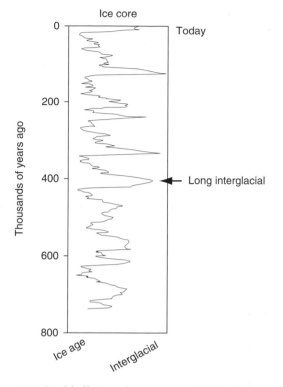

Figure 1.3 Rapid climate changes over the last 800 000 years recorded in Antarctic ice. This sketch is after a section of the figure in a 'News and Views' feature in *Nature*, 'A great grand-daddy of ice cores', by Professor Jerry McManus (2004), marking the publication of an article in that journal by the EPICA consortium (European Project for Ice Coring in Antarctica). The air temperatures over Antarctica are inferred from the ratio of deuterium and hydrogen in the ice at Dome C. The interglacial through which we are now living is thought likely to be of comparable duration to the 'Long interglacial' (more formally known as 'Marine Isotope Stage 11') indicated on this sketch (see also Figure 1.4).

Those same changes in climate deep in geological history have for many years been interpreted differently by others. These past changes have provided the basis for a scepticism that is different from that of Lawson and Lomborg and potentially far more damaging to us all in practice: the scepticism of the oil industry. Few of us involved in exploration and production in recent years can afford to be sanctimonious about the doubt promoted by sections of our industry. Yet although cynics will rapidly point out that it was in our commercial interests to deny the scientific evidence, we did have reasonable as well as unreasonable concern about the rationale behind the attacks mounted on us by the environmentalists.

In the mid-1960s, some time before I came to work full-time in the oil industry, I was a research student at Harvard University, working in the diverse group led by the late Professor Raymond Siever. Siever was a scientist recognised for his rare ability to consider Earth as a whole, in particular as a pioneer of modern global geochemistry. In his classes he introduced us to quantitative analysis of Earth's carbon cycle, but he remained cautious at that stage and for some years thereafter in predicting the effects on climate of the excess anthropogenic carbon dioxide (Press and Siever, 1978, pp. 319–321). Later he became a good deal less cautious. In May 1997 I met Ray and his wife Doris for an elegant English afternoon tea in London. The slightly decadent air of the richly upholstered lounge of the Park Lane Hotel was the perfect setting for the new tale he had to tell me.

That afternoon, as we set about the scones, I proudly recited the key equations I had learnt in his class, alongside Miriam Kastner and Fred Schwab, both of whom were to become influential participants in the climate-change debate decades later (Professor Kastner was a contributor to the 2003 BP–ExxonMobil debate on climate change considered in detail in Chapter 3; see Professor Schwab's (2007) article in *Geotimes* for an example of his pungent approach to the issue). But Siever chastised me for 'characteristically remembering the relevant equations perfectly, except for the critical number giving the maximum rate at which excess carbon dioxide might be absorbed into the ocean'. He had recently made new calculations of global warming in the light of data showing a continuing rapid increase in the atmospheric concentration of carbon dioxide, collected from Hawaii by Professor David Keeling (Figure 1.2). Nodding emphatically over the Earl Grey tea, Siever assured me that we were indeed facing a serious problem.

During the years between these encounters with Siever, much of that time involved with and then embedded in the oil industry, I was not much concerned about climate change. I was

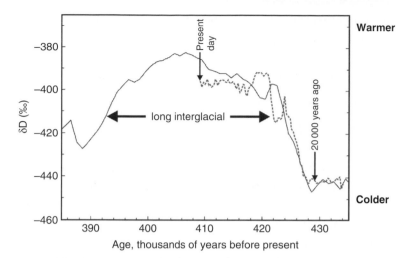

Figure 1.4 This sketch is after Figure 5 in the EPICA (2004) article in *Nature* (lead author Eric Wolff, see Figure 1.1), reporting on the 'great grand-daddy of ice cores' introduced in Figure 1.3. A measure of deuterium against hydrogen is plotted against the vertical axis. The lower the number (upwards on the graph) the higher the inferred temperature. Two ages are compared: the 'Long interglacial' of Figure 1.3, an interglacial 400 000 years ago (continuous line, lower horizontal axis) and the present interglacial (dashed line on same scale). The beguiling similarity of these two curves is being examined closely (Dickson *et al.*, 2009).

reassured by the belief that any global warming caused by our pursuit and use of fossil fuels was at worst postponing an oncoming ice age – or at least 'a long-term trend over the next seven thousand years … toward extensive Northern Hemisphere glaciation' (Hays, Imbrie and Shackleton, 1976, p. 1121). The last crumb of comfort on that score disappeared when I read the results of work on Antarctic ice cores by Dr Eric Wolff and his colleagues (EPICA community members, 2004, p. 623) who concluded: 'our results may imply that without human intervention, a climate similar to the present one would extend well into the future'. Figure 1.4 shows the basis for Wolff's conclusion that we cannot rely on an impending ice age to counterbalance anthropogenic warming.

In the face of the steadily growing evidence from both ice and rocks, even putatively self-centred oil folk can change their minds. But can they change their behaviour too? Does it matter if they don't? For all our sakes the answer to those two questions had better be yes.

1.2 GEOLOGISTS, THE OIL INDUSTRY AND CLIMATE CHANGE

The oil industry has earned a reputation for being sceptical about the hypothesis that significant climate change is caused by the use of fossil fuels. The most obvious reason for this scepticism is self-interest. That unremarkable conclusion does not get us very far. There is another less obvious reason that lies in the educational background and technical expertise of those working in the oil industry.

Experienced, highly trained and scientifically literate geologists and engineers dominate the oil industry. They generally share the most senior positions and they certainly do a great deal of the work to find and produce the oil and gas. Their years of operational experience incline them to scepticism about computer-based models of complicated natural systems: their own forecasts on rocks and fluids and their interaction are constantly tested and at times found wanting. When it comes to considering climate change, they are more likely to be convinced by appeals to geological evidence, such as that from the 55 Ma warming event.

Like many others, I enjoy being part of a discussion that converts a sceptic to your cause. I have instigated conversations on climate change with colleagues in the oil capitals of North America, in Anchorage, Calgary and Houston, in the hope of such an outcome. Some of the best of these discussions have been with my friend Elsworth Boswell in bars in conservative Midland, Texas. Bars in Midland are not necessarily a natural habitat for a former resident of Massachusetts, even when he is fortified by Samuel Adams Boston beer. Boswell, a veteran oil man (and former Little League baseball coach of George W. Bush) was a welcome companion, although our discussions seldom ended with any change of mind or heart. Apart from the fun of the exchange itself, why persist with the argument? What is the practical use of convincing sceptical oil folk that we do indeed have a problem with climate change?

The answer is that these same petroleum geologists and petroleum engineers also have the expertise, possessed by few others, to work together to put large volumes of carbon dioxide into safe underground storage; a process now regarded by some politicians, scientists and oil folk as the ultimate carbon offset. The hard heads in the oil industry are not likely to talk of salvation, but they believe they can offer a solution. They might say: 'If carbon is the problem, we can deal with it. But we shall of course require a fee for doing so. You can pay us twice, once for taking fossil carbon out of the ground, then again for putting it back when you've had the use of it.'

A readiness to 'put it back' requires conviction on the part of a petroleum geologist that this is worth doing for reasons that lie beyond dollars. As noted, the scientific case for anthropogenic climate change has been presented most forcefully by climatologists, rather than by geologists, who have until quite recently tended to be more sceptical. As research into climate change has expanded, so a new breed of Earth scientist has evolved, with the ability to look both backwards and forwards in time. But it remains broadly true that climatologists focus on predictions of the future, relying on a combination of observation of past trends and computer modelling to make their forecasts. Geologists look back in time, using the present as the key to the past. They rely on observation of the geological record of past changes in climate to guide their views on what might happen in future: they are happiest when basing their predictions on the solid ground of rocks. They nurse suspicions of computer models that predict behaviour of planet Earth in the absence of an understanding of all the major controls, even in the face of crisp reassurance by the Royal Society (2007) (Table 1.1), presented in a tastefully green-shaded booklet that sets a standard for effective communication of multidisciplinary science to a wider audience.

So what does constitute really solid evidence for a geologist? As will be set out in some detail in Chapter 2, examination of cores of sediment recovered from drilling into layers of rock beneath the floor of the Atlantic Ocean shows clearly that an episode of global warming occurred 55 million years ago (Norris and Röhl, 1999). Thanks to major advances in dating of rocks

Table 1.1 *The Royal Society gives reassurance about computer modelling of future climate. Here is the answer to 'Misleading Argument 5' in the Royal Society's (2007) brochure:* Climate Change Controversies: A Simple Guide.

Misleading argument: computer models which predict the future climate are unreliable and based on a series of assumptions.

What does the science say?

Modern climate models have become increasingly accurate in reproducing how the real climate works. They are based on our understanding of basic scientific principles, observations of the climate and our understanding of how it functions.

By creating computer simulations of how different components of the climate system – clouds, the Sun, oceans, the living world, pollutants in the atmosphere and so on – behave and interact, scientists have been able to reproduce the overall course of the climate in the last century. Using this understanding of the climate system, scientists are then able to project what is likely to happen in the future, based on various assumptions about human activities.

It is important to note that computer models cannot exactly predict the future, since there are so many unknowns concerning what might happen …

… While climate models are now able to reproduce past and present changes in the global climate rather well, they are not, as yet, sufficiently well-developed to project accurately all the details of the impacts we might see at regional or local levels. They do, however, give us a reliable guide to the direction of future climate change. The reliability also continues to be improved through the use of new techniques and technologies.

---

made during the twentieth century, it is now possible to examine the evidence of that warming episode using a timescale with a resolution of thousands of years rather than the million or so years previously attainable.

The key to this high-definition timescale is the hypothesis advanced by Milankovitch (1941) that regular cycles in Earth's climate are caused by regular variations in the amount of heat received from the Sun. These periodic variations in insolation are related to three aspects of Earth's orbit around the Sun. These three aspects are: (1) changes in the tilt (obliquity) of Earth's axis; (2) variations in the shape (eccentricity) of orbit; and (3) changes in the direction of tilt of Earth's axis at a given point in orbit. The cycles of climate caused by these orbital variations are reflected in variations in the deposition of successive layers of sediment on the floors of oceans and lakes. These layers thereby contain their very own historical markers: 'the astronomical timescale'.

This astronomical timescale, brilliantly developed by Professor Sir Nick Shackleton and his colleagues (Hays, Imbrie and Shackleton, 1976), has given geologists a new high-definition view of the past, beyond the original dreams of those of us raised on the million-year intervals of the fossil-defined classical timescale. Now we can roll up our sleeves and study the implications of this astonishing advance in knowledge. Prime among these implications from a practical viewpoint is that we can now consider the results of the 55 Ma warming event with that new resolution of thousands not millions of years: we thereby read the message in the rocks on a human timescale. The conclusion is remarkable. By natural means the planet has already demonstrated the damaging effects on life on Earth of releasing rapidly a large volume of carbon dioxide into the atmosphere.

Geologists have provided a powerful answer to the sceptics and a motive for their own involvement in tackling the challenge of carbon. Appropriately enough, the oil man who was instrumental in starting the public greening of the oil industry is a geologist: as are others who played a significant part on the road to agreement at Kyoto.

1.3 THE OIL MAN, THE ENVIRONMENTALIST AND KYOTO

The important detailed evidence supporting the significance of the 55 Ma warming event for humankind was not publicly available in the years leading to the Kyoto meeting in 1997. Yet even without that considerable strengthening of their hand, geologists did play an important part in ensuring agreement at Kyoto in 1997. Dr David Jenkins (Figure 1.5), Technical Director of BP, formerly the company's Chief Geologist, wrote a crucial

Figure 1.5 Dr David Jenkins, formerly Chief Geologist, Technical Director of BP, Director of BHP Billiton, now Director of Chartwood Resources. Jenkins wrote the crucial memorandum on 13 January 1997 that led to the historic speech on climate change by John Browne at Stanford University on 19 May 1997. Photograph from David Jenkins.

memorandum to the BP managing directors on 13 January 1997, suggesting that that the company alter its policy on climate change. This proposal was accepted and announced in a speech at Stanford University on 19 May 1997 by BP's Chief Executive Sir John (now Lord) Browne (Figure 1.6).

Sir Mark Moody-Stuart (Figure 1.7), also a geologist, was simultaneously guiding Shell into new territory on climate change – again with significant impact at Kyoto. Before Browne spoke at his alma mater in May 1997, Shell had published its (1996) Annual Report. It included the phrase: 'there is now sufficient scientific evidence [on climate change] to support taking prudent precautionary action.' The significance at Kyoto of that change in attitude was recognised by Dr Jeremy Leggett (Figure 1.8) in his account (1999) of events. Leggett himself drew on considerable geological expertise, having joined Greenpeace after years as lecturer in geology at Imperial College, London and consultant to the oil industry.

Figure 1.6 Lord Browne, formerly Chief Executive Officer of BP, who was instrumental, as was Sir Mark Moody-Stuart of Shell, in leading the change in the attitude of the oil industry towards its release of carbon dioxide to the atmosphere, both in its own operations and through the use of its products by its customers. Photograph taken at the Geological Society, Burlington House, Piccadilly by Dr Ted Nield.

Jenkins, Moody-Stuart and Leggett came to the climate-change issue from a number of different angles and backgrounds, but they had in common Fellowship of the Geological Society of London. Moody-Stuart became President of the Society at a critical stage in the early twenty-first century; Jenkins and Leggett are both Society medalists. The Society has a Dining Club that meets regularly in an airless basement of the Athenaeum Club in Pall Mall, a room formerly used for billiards and to sequester female visitors in that otherwise splendid building. In his account of what he styles *The Carbon War*, Leggett (1999) writes of an encounter with Jenkins at the Athenaeum in 1991 that provided an important thread to Kyoto and beyond.

Leggett had been a favourite son of the Geological Society during the 1980s. At the end of that decade he appeared to have

Figure 1.7 Sir Mark Moody-Stuart, formerly Chairman of Shell and Anglo American. As President of the Geological Society he chaired the debate on *Coping with Climate Change* between Vice President Mr Greg Coleman of BP and Vice President Dr Frank Sprow of ExxonMobil in London on 26 March 2003. Photograph from International Institute for Sustainable Development, by permission of Sir Mark Moody-Stuart.

the academic world at his feet, having won two Society awards. From his position at Imperial College he could have gone on to combine academic geology with oil industry consultancy in a stimulating and potentially lucrative fashion. Instead he chose to set out on the difficult path that was to bring him into regular and bruising contact with those who felt betrayed by his choice. Some of these detractors, not least those connected with the oil industry, were also concerned that Leggett over-interpreted observational science in his environmental work. Not for nothing does the title of his book invoke military metaphor: Leggett claims he chose a life in the trenches with Greenpeace and he did.

At the Geological Society Dining Club in 1991 Leggett was accused by Jenkins of jumping ship. Leggett explained his motivation: the implacable arithmetic of carbon in the atmosphere and in the fossil fuels left below the ground. In *The Carbon War* he

Figure 1.8 Dr Jeremy Leggett, formerly of the Earth Sciences Department at Imperial College London and then a Director of Greenpeace, who is now Chairman of Solar Century. Leggett (1999) has published a vivid account of Kyoto 1997 in which he is kinder to BP and Shell than he is to Exxon and Mobil (then separate companies). Photograph from Jeremy Leggett.

claims that Jenkins made notes on a napkin. That is one of the few errors of fact in Leggett's book that I can prove from personal experience. Nobody would write on the beautifully laundered napkins in the Athenaeum, especially the socially immaculate Jenkins who always carries a notebook in which he records important matters neatly with an expensive pencil. But there is little doubt that he thought his conversation with Leggett worth recording in that notebook.

By the end of the decade Kyoto had proved a triumph for the environmentalists and Leggett himself had moved on from Greenpeace to found a solar energy company. He still has hard words for those involved in frontier exploration for oil and gas.

But he should look back at his dinner with Jenkins with more than the usual satisfaction of those who go through the unchanging Dining Club rituals of science, gossip, port and snuff. For it was Jenkins, several years later, who was instrumental in changing the environmental policy of BP. The memorandum he wrote to the managing directors in January 1997 contains plenty of disputed science, but few could quarrel with the prescient analysis of the position of the oil industry. It is a most unusual business memorandum among the inconsequential millions of jargon-ridden documents produced routinely in large organisations: it is clear, effective and significant way beyond corporate boundaries.

> To: The Managing Directors
> cc: Klaus Kohlhase; Peter Scupholme
> From: Dr David AL Jenkins
> Date: 13th January 1997
> CLIMATE CHANGE
> You've had a pack of information for next week's discussion. I've kept this personal assessment separate as it gives you my view of the solution before we've had the debate and I don't want it to get 'in the way'. As the Technology Advisory Board observed three years ago DALJ is decidedly 'opinionated' on the subject!
> However, as a distilled one pager you may find it a helpful summary.
>
> The past two millennia have been characterised by marked and rapid changes in the climate. Quite unrelated to changes in CO_2 concentration.
> It is certainly now getting warmer. It's been doing so, at least in the northern hemisphere, for the past 200 years, since the end of the 'mini ice age' (c.1400–1800 AD). Based on the cyclicity of the current interglacial we can expect further warming, quite independent of changes in CO_2 levels.
> The current debate is focused on endeavouring to determine the component of climatic change that is now related to the recent rapid rise in anthropogenic CO_2. However, the mean of the rate of temperature change projected for the 21st century, from the IPCC models, is very similar to what was experienced (locally) in the northern hemisphere, in the first 100 years of temperature recovery from the 'mini ice age'. It is also close to what we've experienced (as a global average) during periods of the 20th century.

Based on this experience, then for humans, adaptation to a warmer world really should be feasible and primarily an issue of more effective management of water supplies. The practical concern on migration of faunal/floral boundaries, will be related to those areas which move towards a frost free environment.

Sea level rise is almost a red herring. *[This is an example of the 'disputed science' to which I refer above; sea-level changes are discussed in Chapter 2 of this book.]* The Maldives were there in 1150 AD when northern Europe was then as warm as the IPCC project it to be in the second half of the 21st Century. And the climate models predict increased precipitation on the ice caps.

However whilst a warming world continues to be a political concern, the hydrocarbon industry should be prepared to respond in a constructive manner.

One way would be to consider promoting the concept of CO_2 sequestration, separating and putting the combustion gases from our fields and refineries back under the ground. Large fixed source emitters put out over half the anthropogenic CO_2. The technology exists to capture and sequester these gases, if society wishes to pay the increased cost of energy in so doing.

But the real point of shifting the debate is it puts us on a playing field with the power sector. The implication would be that they could and should respond also. It would also address the vexed issue of rapidly rising emissions in the developing countries and provide a vehicle for effective tradable permits.

If the debate moved this way, it could divert the focus from the use of hydrocarbons as transport fuels. This would certainly help the longer term transitioning to alternative sources. It might also help achieve more clarity in the purpose and effect of fiscal incentives and disincentives.

With this stance we could also then press more credibly for attention to adaptation. This will certainly be needed, regardless of future trends in CO_2 emissions and hence is the socially responsible path along which to direct effort and investment. The likely outcome will be a compromise, but at least we'd be having the discussions in a full carbon cycle context, rather than just focusing on emissions.

BP's change in policy is quite properly given prominence in Leggett's (1999) account, as is the impact of environmentally converted oil companies such as Shell in the vital green triumph at Kyoto. That triumph would not have happened without the information and pressure applied to the industry by Leggett and his colleagues in the environmental movement. Nor would it have happened without fellow geologists and others in industry listening carefully and then thinking for themselves, true servants of their companies rather than being mere peddlers of corporate convention. In the geological community at large, ferocious arguments on matters to do with planet Earth frequently become wonderfully and cheerfully personal and hence real in their impact, yet ephemeral in their effect on the comradeship. In the case of climate change, some tension between geologists may remain, for this is not just a dispute about rocks.

I came by my own copy of Leggett's book at the Sir Peter Kent Lecture in Burlington House given by the Rt. Hon. Michael Meacher (then Minister of State for the Environment). The date was 24 November 1999, just after the publication in *Nature* in October of the crucial paper by Norris and Röhl on the 55 Ma warming event. Discussion following Meacher's lecture included the new confidence we now had in dividing geological time into thousands rather than millions of years and its application to reading on a human timescale what is written in the rocks. There was Leggett in the leathery Lower Library of the Geological Society, his hand on a pile of *The Carbon War*, with the eyes of a man who had been in a fight, if not exactly a war.

We discussed this most recent geological vindication of the predominantly climatological arguments he had deployed in his pre-Kyoto campaigning. He thought the geological line of attack was unlikely to mollify those amongst his former academic colleagues who felt that he had stretched the scientific evidence in making his point during the 1990s. He signed my copy of *The Carbon War*, adding a cheerful reference to past and future discussions. Here was a man not pleased to be right that climate change was a problem, but feeling vindicated that he had made such an effective fuss.

The important message interpreted from the rocks by Norris and Röhl in 1999 was outstanding confirmation of earlier work by Dickens, Castillo and Walker (1997), and Dickens, O'Neill, Rea and Owen (1995) (see Chapter 2). If a broader public had been aware of that message in 1997, would that have prevented the rapid development of a major Atlantic Divide in the oil industry? What we do know is that after Kyoto, common ground on climate change tended to run out somewhere near Iceland. The lines were drawn so that the two largest European oil companies began to act on the belief that mankind is exerting a measurable influence on Earth's climate, while in the USA, leading figures within the industry continued to incline to a belief that any present or future changes in climate may be ascribed to non-anthropogenic causes. The public and politically important changing of minds in Europe had led to public and strenuous argument across the Atlantic: the oil industry found itself challenged and divided by carbon.

Yet even as the divide developed, new evidence from the rocks was offering the path to agreement and progress: the detailed history of the Paleocene–Eocene Thermal Maximum – the 55 Ma warming event. In the next chapter that warming event will be examined, before we return to the opening and closing of the Atlantic Divide in Big Oil in Chapter 3.

2

A crucial message from 55 million years ago

One of the best-documented examples of climate change preserved in the geological record occurred 55 million years ago (55 Ma). The changes it caused to life on the planet mark the boundary between the Paleocene and Eocene epochs. The Paleocene–Eocene Thermal Maximum (PETM) or Latest Paleocene Thermal Maximum (LPTM) was a dramatic global warming event that has a clear bearing on present-day affairs: its siren message challenges those still sceptical of our ability to change Earth's climate.

Professor Gerald Dickens and his colleagues (1995, 1997), introduced in Chapter 1, reach several conclusions from their studies of the 55 Ma event that are of particular significance to our current concerns about anthropogenic climate change. The first of their conclusions is that a large quantity of carbon was released into the ocean–atmosphere system over the geologically short period of some 10 000 years. The second observation is that the temperature of the water at the bottom of the ocean increased rapidly by more than 4 °C, from 11 to 15 °C, over the same short period. The third conclusion is that the oceans became notably more acidic. All this was accompanied by a general and significant global warming. Ominously from our contemporary perspective, Dickens concludes that it took some 200 000 years for the planet to return to something resembling the conditions prevailing before the massive and sudden release of carbon took place.

Detailed confirmation of the conclusions of Dickens and others comes from Norris and Röhl (1999). They consider carbon cycling and the history of climate warming 55 million years ago using late twentieth-century techniques for dividing geological time into thousands of years, rather than the classical millions of years based on the fossil record of evolution of life (see Chapter 1). Their observations are here summarised in Figures 2.1 and 2.2. Professor Richard Norris and Dr Ursula Röhl suggest that a massive volume of carbon was released abruptly from the pores of rocks beneath the sea floor, in which it had previously been trapped (see Figure 2.1). This release pumped carbon into the atmosphere at rates similar to those induced today by human activity. Fifty-five million years ago there were profound effects on the oceans and atmosphere.

We can study these effects by a painstaking analysis of 55-million-year-old sedimentary rocks, formed particle by particle, layer by layer, on the floors of oceans and lakes. Cores of these layers of sedimentary rocks can be obtained for analysis by drilling into the rocks lying beneath the floor of present-day lakes and oceans. The fossils of animals and plants contained in these layers tell us a lot about the environment in which they developed, including changes in climate recorded progressively through time. The composition of the shells of marine fossils gives import-ant insights into the nature of the ocean waters from which they formed; for example, relatively high ratios of magnesium to cal-cium in shells made of carbonate show a rise in temperature of the oceans of 4–5 °C at 55 Ma (Tripati and Elderfield, 2005).

Layers of different composition and fossil content were laid down before, during and after the 55 Ma warming event, one on top of another, each reflecting changing aspects of Earth's environment. Reading what is written in those rocks is compara-ble to enjoying a history such as Foster's (1988) account of *Modern Ireland*, which runs in classical style from the seventeenth to twentieth centuries. So, as with Foster, you start with the oldest layer, commonly at the bottom of a pile, and then work upwards through successively younger rocks.

By studying the rocks in this fashion, we can say with some confidence that the sedimentary record at 55 Ma indicates that excess carbon remained in the atmosphere for about 170 000 years

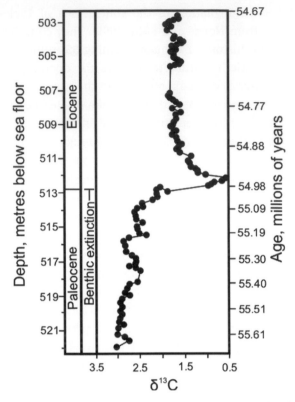

Figure 2.1 Evidence for a large and rapid release of carbon to the atmosphere 55 million years ago, after Norris and Röhl (1999, their Figure 2), read from a core taken beneath the deep-sea floor of the western North Atlantic Ocean, at Ocean Drilling Programme (ODP) Site 1051 in 1997 (see Figure 7.3). The carbon released was markedly rich in the carbon-12 isotope relative to the carbon-13 isotope (horizontal axis). Living cells preferentially incorporate carbon-12 and hence their fossil remains are relatively poor in carbon-13. We may therefore infer that the carbon-13-poor carbon released at 55 Ma came from 'fossil' carbon, such as that stored in the subsurface in methane hydrates or ancient sedimentary rocks rich in organic carbon. Geological time is plotted (on the right-hand vertical axis) with a definition of thousands of years, rather than the traditional millions of years used by geologists until recently. This greatly improved definition of time is made possible by the technique illustrated in Figure 2.2. The abrupt change in the fossil record ('Benthic extinction' – 'benthos' are organisms that live on or near the bed of the sea) and other properties of the rock can thereby be placed at 54.98 million years.

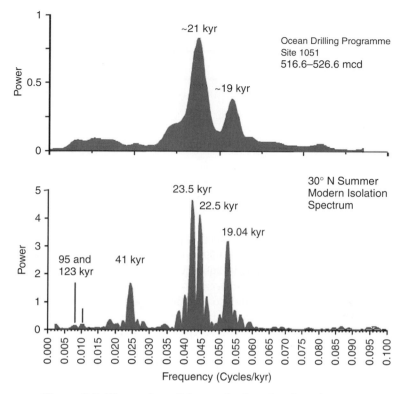

Figure 2.2 Illustration of the method used to date the 55 Ma warming event (Norris and Röhl, 1999), based on graphs provided by Professor Richard Norris (November 2008). The upper graph illustrates the dominant frequencies of periodic variations in climate-related properties of sediments in a core cut over 500 metres below the sea floor ('mcd' in upper graph) at Ocean Drilling Programme (ODP) Site 1051. These variations in properties of the sediments in the core are matched in the lower graph with variations in modern summer insolation at 30° N. We know the timings of climate changes related to variations in Earth's orbit, so we can plot the 55 Ma event on a human timescale (kyr, 1000 years)

(Röhl, Westerhold, Bralower and Zachos, 2007), much as suggested previously by Dickens and others. We can also say with some confidence that there is a striking correspondence between the rate at which large volumes of carbon were introduced without our influence 55 million years ago, and the rate at which large volumes are now being put into the atmosphere by us (Figure 2.3, see also Figures 1.1 and 1.2).

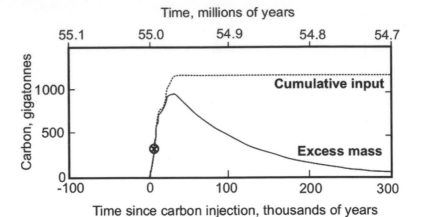

Figure 2.3 Sketch after a section of an illustration used by Professor Gerald Dickens (1999) to show the rapid release 55 million years ago of carbon and its subsequent removal (see 'Excess mass') from the atmosphere and oceans over 100 000 years. We may use this 55 Ma event as a guide to the effect of our present-day release of carbon, should this remain unchecked. We have so far climbed at least 300 megatonnes up the steep slope of 'Excess mass' that begins at zero (55 Ma), so we have already reached at least as far up the curve as the point marked X (see also Figures 1.1 and 1.2).

Norris and Röhl published their 1999 paper in *Nature*. Each week in that journal selected papers are discussed in a separate section by a fellow-expert. In this case the additional note came from Dickens (1999), who as we have seen is a leader of research into the 55 Ma warming event. In 'Carbon cycle: the blast in the past' he is emphatic (p. 752) about the significance of the new data:

> On current estimates, over a period of less than a thousand years 2000–4000 gigatonnes of carbon will be added to the atmosphere by human activity. That's 2–4 million million tonnes. What will be the consequence of this rapid and massive release of carbon? The question has been tackled primarily with numerical simulations of the global carbon cycle constrained by experiments, present-day observations and records from the late Quaternary, the past 200,000 years or so of Earth history.
>
> An alternative – studying ancient blasts of carbon – has always seemed pointless simply because we thought that there weren't any such blasts; as many of us know, natural processes

cannot suddenly add enormous amounts of carbon to the ocean or atmosphere. That view of the global carbon cycle is spectacularly flawed, however, as highlighted in the paper by Norris and Röhl on page 775 of this issue.

For just a brief period, about 55 million years ago, temperatures at high latitudes and in the deep oceans soared by 5–7 degrees C. This event, called the late Palaeocene thermal maximum or LPTM, coincided with an extraordinary decrease in the 13C/12C ratio (delta13C) [see Figure 2.1] of … carbon on the Earth's surface.

Figure 2.3 is from Dickens' essay in *Nature*. Through our own activities we have climbed at least 300 gigatonnes (Gautier, 2008) – over 500 gigatonnes according to Schmidt and Archer (2009) – up the steep slope of the curve starting at zero years, the time of the 55 Ma injection of the carbon on the Dickens graph shown in Figure 2.3. That anthropogenic climb began markedly in the mid-eighteenth century and the curve has recently steepened: 'Half of the total emissions have occurred since the mid-1970s' (Gautier, 2008, p. 78). Our present rate of release of carbon to the atmosphere is currently over 9 gigatonnes each year (Schmidt and Archer, 2009), with some 80% of that caused by burning fossil fuels (Gautier, 2008). Given a set of reasonable assumptions about similar behaviour of the present-day and the 55 Ma carbon cycles, we can see that we will leave an imprint in the ocean, atmosphere and biomass for well over 100 000 years. No wonder Dickens concludes that: 'With a revised model for the global carbon cycle, and a single, well-dated late Palaeocene section, we can now begin to view aspects of Earth's future in an entirely new light.'

Unsurprisingly the 55 Ma event has been studied with increasing intensity in recent years and there is now a voluminous and fascinating literature based on observations made on rocks of that age exposed on the present-day land surface and recovered by drilling through sediments beneath the sea floor. Nothing has been discovered in the last decade that brings into serious question the main conclusions reached by Norris and Röhl (see Cohen, Coe and Kemp, 2007), although significant new information continues to be published. For example, Handley, Pearson, McMillan and Pancost (2008) have presented observations from sediments in Tanzania that bear on the question: just how much carbon was released 55 million years ago? This topic is discussed further later in this chapter: the salient point here is the similarity to present-day inputs of carbon by us.

The effect on life in the oceans of the 55 Ma event is particularly well documented and confirms the earlier conclusion of Thomas and Shackleton (1996, p. 401) that at that time 'deep sea benthic foraminifera [organisms living at or near the bed of the sea] suffered their only global extinction of the last 75 million years and diversity decreased worldwide by 30–50% in a few thousand years'. As a result of the massive injection of carbon dioxide into the ocean–atmosphere system, the oceans became warmer and more acidic. There was a marked decrease in the thickness of the habitually less acidic surface layer of the oceans, in which organisms can build their shells with calcium carbonate; in deeper waters calcium carbonate dissolves. A boundary between the less-acidic surface waters and the more-acidic deeper waters is defined by the depth below which calcium carbonate dissolves, termed the Carbonate Compensation Depth (CCD). At 55 Ma the CCD moved closer to the surface of the sea.

As the water warmed up 55 million years ago it expanded. Thomas and Shackleton (1996, p. 420) state that an ocean-wide increase in temperature of high latitude surface waters and deep waters worldwide, by 4–6 °C 'would have the effect of raising sea level by about 5–6 m'. The Paleocene climate was already warm and the planet had few if any significant permanent accumulations of ice even before the 55 Ma warming event. In cooler times such as our own a comparable warming event would lead to a far greater rise in overall global sea level as land-based ice melted.

The effects of the 55 Ma warming event onshore are manifest in the 'large number of extinctions in mammal groups that had been dominant in the Palaeocene …' (Hooker, 1996, p. 205). Hooker claims this time of extinction as 'One of the most important events in mammalian history during the Cenozoic [last 65 million years], and certainly the most important within the northern hemisphere Paleogene [65 Ma to 23 Ma].' Three modern orders of mammals, Artiodactyla, Perissodactyla and Primates, first appear in the fossil record at the Paleocene–Eocene boundary, at 55 Ma.

'Their appearance on all three northern continents has been linked to diversification and dispersal in response to rapid environmental change at the beginning of a worldwide 100,000 – 200,000-year Paleocene–Eocene thermal maximum (PETM) and carbon isotope excursion. ... global environmental events such as the PETM have had profound effects on evolution in the geological past ...' (Gingerich, 2006, p. 246).

A latter-day, conscious and arguably rather specialised mammal such as *Homo sapiens* might conclude that the Paleocene–Eocene Thermal Maximum was the type of change in climate that it would be better not to precipitate through one's own agency. A sensible reading of the message written in the rocks is that it would be best to simply relish the contingency that led to that distant primate ancestor of our species, and be determined not to repeat it for the benefit of some unknown creature 55 million years hence.

That cautionary tale seems to be based on science that is clear enough. If a significant amount of carbon is added to the atmosphere, in the form of methane or carbon dioxide or both, the result will be a warming of the surface of the planet. There is a little twist to that story, much beloved of the climate-change sceptics. This arises from an observation made from cores of ice taken from Greenland and Antarctica, recording events within the last million years. At the beginning of the periodic warming events associated with the variations in Earth's orbit around the Sun (see Chapter 1), a rise in temperature may precede an increase in the atmospheric concentration of carbon dioxide – rather than the other way round. The sceptics should not draw much comfort from this. If the oceans are being warmed by a periodic increase in heat coming from the Sun, a predictable consequence of that will be the release of increasing amounts of carbon dioxide from the oceans.

Following that periodic warming arising from orbital (astronomical) causes, the carbon dioxide appears to take over its normal dominant role, in an intriguing series of events. 'The system is non-linear with feedbacks' (Professor Harry Elderfield of Cambridge University, personal communication, 2009):

> It could well be that a tiny amount of orbital forcing (insolation increase) gets things going and that carbon dioxide then drives

temperature. We know that orbital changes do not add sufficient solar energy to Earth's surface to explain the interglacial
temperature increase from glacial times. That change is because
of carbon dioxide increase.

As demonstrated in Figures 2.1 and 2.2, it is the observation of
regular, periodic, climatic events in the geological record that
enables us to bring the irregular climatic events like the 55 Ma
warming onto a human timescale. Because the periodic climatic
events are ultimately controlled by variations in Earth's orbit,
they are a regular and persistent feature of our planet and there
is bound to be interaction between periodic events and any given
episodic event, whenever that episodic event arrives. That interaction is not yet fully deciphered in the geological record, but that
does not mean we should postpone delivery to the world at large
of a crucial message from that record.

The crucial message recorded in ice and rocks is that release
of carbon to the atmosphere, at the rate now practiced by us, is
taking us dangerously beyond the range of the geologically recent
periodic increases in atmospheric carbon dioxide recorded in the
ice cores. We can establish links between changing climate and
changing concentration of carbon dioxide in the atmosphere. We
know that early members of our species survived these periodic
changes in climate, maybe even benefited from them. Could we
survive even these smaller periodic variations now, in our specialised billions, without desolation? Still more daunting, could we
cope with something episodic and huge, on the scale of a 55 Ma
warming event?

2.2   WHAT CAUSED THE 55 MA EVENT?

The short answer is that we are not yet certain of the exact
nature of the trigger or triggers, although there are plenty of
scientists hot on the trail (for example, Lourens *et al.*, 2005;
Panchuk, Ridgwell and Kemp, 2008). This uncertainty is scientifically fascinating, but, as we shall see in the following section (2.3)
does not affect critically the warning message to us from 55
million years ago. First let us consider the current debate on the
trigger of the 55 Ma event.

An early and still plausible candidate as trigger is a geologic-ally sudden release of methane that had previously been trapped in large volumes in gas hydrates formed in sediments below the ocean floor. This explanation is favoured by Cohen, Coe and Kemp (2007): they also suggest this mechanism as a trigger for a comparable older (Early Jurassic, 180 Ma) episode of global warm-ing (Hesselbo *et al.*, 2000; Kemp, Coe, Cohen and Schwark, 2005). Gas hydrates are solids, resembling ice. They are formed of a mix-ture of water and gas that is stable at high pressure, low temperature and high concentration of gas. So these hydrates become unstable if pressure is reduced or if temperature is increased.

What could trigger instability in the methane hydrates? One hypothesis, favoured by Cohen, Coe and Kemp (2007), is that this instability may have been primed by an earlier, episodic natural release of carbon that led to some warming. A subsequent major release of methane from hydrates was then triggered by a periodic increase in heat reaching Earth from the Sun at a particular point in Earth's orbit. In the case of the 180 Ma Early Jurassic warming event, Cohen and colleagues suggest that volcanism was the orig-inal, episodic, culprit that primed the system for subsequent release of methane from hydrates – this later release being triggered by a periodic astronomical (Milankovitch) event (see Chapter 1). Whatever the relative timings might be, it is clear that episodic triggers such as volcanism and periodic variations in Earth's orbit can act in sequence to precipitate a major global warming event.

Comparable notions of a trigger by episodic volcanism are also invoked for the 55 Ma warming event, but another possible cause has been suggested (Maclennan and Jones, 2006). Methane hydrates may have been destabilised directly by an episodic event connected with convection in Earth's mantle. The idea is that the 55 Ma warming event was triggered by a transitory uplift of methane hydrates beneath the sea floor, caused by the passage deep below of a hot blob carried along by mantle convection: a notion examined in more detail below. No matter which way the debate on these triggers may eventually turn, it is already clear that the key lies in understanding a particular interaction. This crucial interaction is between the effects at Earth's surface of the periodic variations in heat input from the Sun, and the episodic effects of irregular manifestations of Earth's internal heat, such as volcanism.

The global numbers involved in considering any controls of climate change are beguiling. At first blush they suggest that the primary controlling factor on variations in Earth's climate must be variations in heat from the Sun, not variations in flow of heat from Earth's interior or other Earth-based phenomena. The annual loss of heat from Earth's interior was recognised by Press and Siever (1978, p. 331) to be an order of magnitude larger than the rate at which heat is released from man's own activities. But, in turn, the input of heat from Sun to Earth dwarfs the heat from Earth's interior. That input of heat from the Sun averages 200 watts per square metre, over 2000 times the less-than one-tenth of a watt per square metre loss of heat from Earth's interior (Schiermeier, Tollefson, Scully, Witze and Morton, 2008). The balance of heat at Earth's surface is in part mediated by a gas, carbon dioxide, the concentration of which in Earth's atmosphere is measured in parts per million.

Many important geological controls of atmospheric chemistry and physics had been identified long before the recent attention to past warming events such as that 55 million years ago. For example, it has long been known that carbon dioxide is removed from the atmosphere as the weak carbonic acid in rainfall reacts with rocks exposed on Earth's surface, trapping the carbon dioxide. It is also increasingly well established that there is an interaction between the solid land surface of the planet and the atmosphere above. The Himalayas result from the collision of India with Asia, caused by convection in Earth's mantle that in turn is caused by Earth's internal heat. The mountains have an effect on atmospheric circulation and the atmosphere in turn has an effect on the weathering, erosion and development of the mountains. Press and Siever (1978, p. 339) describe the earlier classical view that is being refined by current research: 'in a real sense Earth's internal heat builds mountains, and external heat from the Sun destroys them'.

On one matter at least the oil industry has continued to agree since 1997: major changes in the climate of Earth have taken place throughout geological time. These changes have, until recently, obviously not been caused by us, although the most recent have affected primate evolution. Some of the changes in climate are related to the effects of heat escaping from inside the planet and are therefore particularly convincing to geologists. Cane and Molnar (2001) suggest that the convergence of Australia and Asia has restricted the passage of relatively warm water from the Pacific into the Indian Ocean. They propose that the emergence from the sea 3 to 4 million years ago of the Indonesian island of Halmahera, may have been responsible for diverting major ocean currents and hence increasing aridity in East Africa, with consequent changes in primate habitat (see also Smith and Pickering, 2003). The main configurations of land masses invoked in such hypotheses result from the movement of tectonic plates, driven by convection in Earth's mantle and ultimately by Earth's internal heat.

Another example of control of climate change by Earth's internal heat comes from the Atlantic Ocean. It is well established that parts of the sea floor underlain by relatively hot rocks are higher than areas of the sea floor underlain by relatively cool rocks. So the Mid-Atlantic Ridge, which is underlain by relatively hot rock where new crust is created by upwelling magma, forms a submarine mountain chain, marking a plate boundary. At Iceland this ridge emerges from the ocean, because there it coincides with a well-known 'hotspot' or 'mantle plume'. (The different names reflect an academic dispute that goes beyond mere terminology but need not detain us here. I do favour the evidence supporting the existence of 'mantle plumes', but choose to use 'hotspot' in this book for general clarity.)

Iceland has long been a source of field observation for geologists and of natural hot springs in which they can relax following their day's work – and of course continue to debate hypotheses concerning the origin of the hotspot/mantle plume. We do now understand that the Iceland hotspot varies in temperature, and hence elevation, over episodes of a few million years. As the temperature of the Iceland hotspot has varied, so has the temperature and hence elevation of a submarine ridge running west–east through Iceland from Greenland to the Faroe Islands and Scotland, at right angles to the tectonic plate boundary marked by the north–south-trending Mid-Atlantic Ridge.

Wright and Miller (1996) first suggested that changes in the elevation of that Greenland–Iceland–Scotland submarine ridge over millions of years had controlled ocean circulation to such an extent that global changes in climate resulted. At times of greater heat in the Iceland hotspot, the ridge is raised like a lock gate, much reducing the deep southward return flow of water brought north as surface flow by the Gulf Stream. Wright and Miller argue that this raising of the lock gate causes episodic cooling of North Atlantic climate. This cooling happens because raising the lock gate blocks much of the 'normal' deep flow southwards across the now-elevated Greenland–Iceland–Scotland ridge, leading to failure of the warm Gulf Stream as the 'conveyor system' of global ocean currents is stalled.

Dr Heather Poore came to assess this hypothesis of Wright and Miller through a series of encounters that typify the ecumenism of present-day Earth science. The chain of events was as follows. In May 1995 I was a member of staff at BP Exploration when Dr Nicky White (Figure 2.4) of Bullard Laboratories, Department of Earth Sciences, Cambridge University, led a seminar for a BP group in Aberdeen. I was in BP's office that day on other business and attended his talk. I did so at the suggestion of Dr David Latin, a former research colleague of White and one of the first recruits into a new BP early development scheme (*Challenge*) that I had been instrumental in establishing in 1993. Latin had since then followed probably superfluous advice from me to keep in good touch with his university roots.

White, like Latin a man of wit and vigour, spoke on a singular heating event and uplift of Earth's surface some 60 million years ago. The evidence presented by White showed that this event affected the region now occupied by Scotland, and was caused by a particularly hot episode in the early development of the Iceland hotspot (see also Chapter 8). In discussion following his talk, I suggested to him that BP's studies of rocks of that age in the North Sea, including Forties Formation, indicated multiple heating and uplift events rather than just one. White and I then collaborated on research that led to a joint Cambridge University–BP Exploration publication (White and Lovell, 1997). That 1997 paper in *Nature* includes a reference to the work of Wright and

Figure 2.4 Dr Nicky White (on the right), Reader in Earth Sciences at the University of Cambridge, receiving the Bigsby Medal of the Geological Society from President Lord Oxburgh in 2001 (see Chapter 5). Photograph taken at Burlington House, Piccadilly, by Ted Nield.

Miller. This was thanks to another unscheduled conversation, this time between White and his then Cambridge colleague, bio-geochemist Dr Ros Rickaby, who was already aware of the then-emerging work of Wright and Miller. White was keen to pursue the Wright–Miller line of inquiry and Poore accepted his invitation to carry out doctoral studies on this topic.

What she found was clear evidence that the Greenland–Iceland–Scotland ridge had indeed acted as a lock gate, at least over the last 12 million years, and had influenced ocean circulation and hence climate (Poore, Samworth, White, Jones and McCave, 2006) (Figure 2.5). Here was a convincing example of the interaction between the effects of flow of heat to the surface of the planet coming from below and that coming from the Sun. The effects on climate were episodic, reflecting times of greater and lesser heat reaching Earth's surface via the Iceland hotspot,

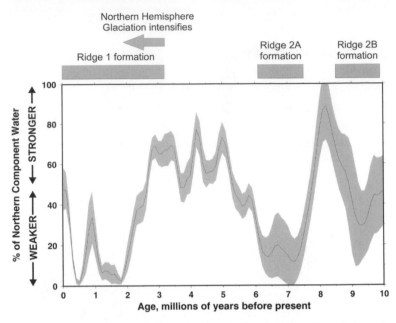

Figure 2.5 Control of North Atlantic flow by episodic variations in the Iceland 'hotspot' (or 'mantle plume'), illustrated by a much-simplified sketch after Poore *et al.* (2006, their Figure 10). The strength of the deep return flow of cold water from north to south ('Northern Component Water' on vertical axis) is plotted against age on the lower horizontal axis (grey shading shows ± one standard deviation in the data), and inferred even hotter pulses in the already hot Iceland mantle plume ('Ridge formation') are plotted on the upper horizontal axis. The elevation of the 'lock gate' running from Greenland through Iceland to Scotland at times of ridge formation appears to have a major influence on ocean circulation and hence climate. This is a recently discovered and important control on climate by Earth's internal heat.

rather than the periodicity of input of heat from the Sun, reflecting predictable and regular variations in Earth's orbit.

The hot pulses in the Iceland hotspot can be traced back to the time of the opening of the northern North Atlantic Ocean some 60 million years ago. A particular pulse in the hotspot 55 million years ago may well have triggered the Paleocene–Eocene Thermal Maximum. Two possible mechanisms related to that hot pulse have been proposed. Maclennan and Jones (2006) suggest that uplift of the sea floor of the early North Atlantic destabilised methane hydrates. Svensen *et al.* (2004) propose a different source for the vast release of fossil carbon: the carbon-rich sediments flanking the ocean into which great quantities of hot magma were intruded, thereby releasing the carbon. In both hypotheses activity of the Iceland hotspot is the primary cause. Both mechanisms could also have acted as the Cohen-style priming mechanism previously discussed. For the purposes of the arguments presented in this book this uncertainty is not critical: the main message bearing on our current concerns about climate change that may be read in those 55 Ma rocks is ominously clear.

There is a nice irony in the North Atlantic story that goes back to the exchanges between White and me at the seminar in BP's office in Aberdeen in May 1995. As a result of that meeting he and I began a period of joint research that continues to this day (Chapter 8). We and our colleagues are trying to reconcile two quite different sets of data, each giving insights into the geological history of the North Atlantic region over the last 60 million years. One set of data is the history of the Iceland hotspot, or 'mantle plume', recorded directly in the igneous rocks formed by activity of that hotspot and studied over many years by academics bereft of commercial motive. The hotter the Iceland hotspot, the greater the quantity of igneous rock formed for the university folk to study. The second record is that preserved in the sedimentary rocks in the North Sea and west of Shetland, derived largely from the erosion of early Scotland. These data were collected by oil folk bent on profit. The Iceland hotspot was then nearer to Scotland than it is now, and was having a big effect in that region. The hotter the hotspot the more volcanic activity there was – and the greater the uplift, exposure and height of the early Scottish land mass. As that land mass was lifted up, large volumes of sandy sediment were poured into the sea on its flanks, not least into the early North Sea. The age of the sandstones of the Forties Formation oil reservoir is 55 Ma.

The research published by Dr John Maclennan and Dr Stephen Jones (2006), which was introduced earlier in this section, arose from their involvement with the White and Lovell research on the pulsing of the Iceland hotspot. Jones was one of our earliest jointly supervised research students. Maclennan's main line of research lay elsewhere, in collaboration with

Professor Dan McKenzie, but his great intellectual curiosity, primarily expressed to me over lunches rich in carbohydrate in the Cavendish canteen, led to collaboration first with me and then with Jones. The linking by Maclennan and Jones of the behaviour of the Iceland hotspot with the 55 Ma warming event was supported by later quantification of the uplift of the Scottish region at that time by Dr Max Shaw Champion (Shaw Champion, White, Jones and Lovell, 2008) and the development of a model of a particular aspect of mantle convection that provides a plausible mechanism for that uplift (Rudge, Shaw Champion, White, McKenzie and Lovell, 2008) (Chapter 8).

By that academic route, a circle of university and industry activity is closed. Crucially supported by data acquired at a cost of hundreds of millions of dollars by the oil industry, and made freely available to us at Cambridge University without any strings attached, a group of young researchers has established a series of links between the Iceland hotspot, the reservoir sandstones of the North Sea, the circulation of the Atlantic Ocean, and the 55 Ma warming event that is so important to our present concerns about climate change. It appears that the uplift event that led to the formation of a body of sand that would later play host to 4 billion barrels of oil at Forties, also precipitated an episode of dramatic climate change. That 55 Ma event provides a warning to us now that uncontrolled release of fossil carbon from Forties Formation oil is not such an unmixed blessing as we thought at the time of its discovery in 1970.

The Maclennan and Jones research has generated new angles of attack on the causes of the 55 Ma warming event. Nisbet *et al.* (2009) suggest that the trigger was release of methane and carbon dioxide at 55 Ma from the Kilda basin, west of Paleogene Scotland, particularly from lakes they compare to those in present-day East Africa. The continuing uncertainty over the detail of causes is the very stuff of academic research, but it can provide sceptics with ammunition and make others nervous about drawing lessons from the past to guide us in coping with present-day concerns about climate change.

Dr Anthony Cohen of the Open University shows no obvious sign of such nerves as he spells out clearly the conclusions of research into the origins of earlier, episodic major warming events in Earth history, both 55 million years ago and 180 million years ago (Cohen, personal communication, 2008):

> It seems to me that the danger of the present-day situation is that we have bypassed entirely the c.80–100,000 year lead-in that occurred at the PETM [55 Ma] and in the Toarcian [Early Jurassic, 180 Ma] because the eight gigatonnes of carbon that is being released each year takes us straight in to the regime associated the very abrupt changes [seen in the geological record] ... For the first time in the history of the planet, we – humankind – have become responsible for releasing huge quantities of additional fossil carbon to the atmosphere–ocean system at an alarmingly high rate. And what makes this even more worrying is that the positive feedback of methane hydrate hasn't yet set in.

This is a cause for alarm to be felt by all, not just by the coal and oil companies that take fossil carbon out of the ground, but also by all those who find their products so useful. Is the 55 Ma event really that convincing? Yes, I believe so. Will Big Oil truly be alarmed? I don't know, but I do know that for the oil industry a message that comes from the rocks comes from a trusted source.

2.3   ALARM FOR BIG OIL FROM A TRUSTED SOURCE

Even for those working in the very centre of the sceptical North American oil industry, the new (Norris and Röhl, 1999) message to be read from the rocks was bound to be somewhat alarming: it came from the geological heart of their business. Geologists knew already that stability was not an option; climate had changed often in the past, in some cases abruptly (Figure 1.3). For many that earlier, familiar message from the rocks had been a reassurance that puny man was unlikely to cause much harm relative to 'natural' Earth processes; with the new results from the deep-sea drilling – drilling bent on science rather than oil – this comfort was now no longer available. It was beginning to look as though the climatologists were right. There is such a thing as anthropogenic climate change and it is not a trivial matter.

For some oil folk and their geological colleagues outside the industry, scepticism would continue to be the order of the day. The published exchanges between the sceptical Dr Lee Gerhard (2004, 2006) and the much less sceptical Lovell (2006) exemplify the continuing debate. Arguments by analogy cannot in themselves constitute sure proof. The geography of the 55 Ma world was different from that of today. For one thing, the northern North Atlantic was only just beginning to open. The climate of the planet preceding the 55 Ma event was already significantly warmer than that of present-day Earth. There was little if any permanent ice. And are we confident that critical elements of the carbon cycle at 55 Ma were then as now?

This boils down to how much confidence we can have in the figures for release of carbon at 55 Ma featured in Figure 2.3. This hinges on the question of the sensitivity of Earth's climate at 55 Ma to increases in the atmospheric concentration of carbon dioxide. 'Climate sensitivity' is a term commonly used, not least by the Intergovernmental Panel on Climate Change. It is defined as the increase in temperature resulting from a doubling in the concentration of carbon dioxide in the atmosphere. This increase in temperature ranges from 1.5 to 4.5 °C per doubling of carbon dioxide concentration.

Pagani, Caldeira, Archer and Zachos (2006, p. 1556) state that 'the Paleocene–Eocene Thermal Maximum (PETM) may be the best ancient analog for future increases in atmospheric CO_2'. They then ask 'How well do we understand it?' They argue that unless global temperature sensitivity at 55 Ma was much higher than currently assumed, even larger amounts of injected carbon were required to generate the recorded high temperatures than the huge volumes suggested by Dickens. In that case methane hydrates could not be the only source of that carbon – there was not enough carbon trapped in the 55-million-year-old methane hydrates to provide the vast quantities required to raise the temperatures by the amount reliably recorded in the rocks. Also in that case, the 300 gigatonnes of carbon we have so far added to the atmosphere would be a smaller proportion of the total release of carbon responsible for the 55 Ma warming event than is shown in Figure 2.3, a smidgeon of comfort in a still bleak story for us.

So what was the climate sensitivity at 55 Ma? Goodwin, Williams, Ridgway and Follows (2009, p. 148) consider climate sensitivity over the last 400 million years. They conclude that the current Earth system is more sensitive to carbon perturbations now, and during the most recent Cenozoic era which began 65 million years ago, than it has been over much of the preceding 400 million years. So, crucially, they believe that present-day climate sensitivity is indeed comparable to the sensitivity at 55 Ma, in the early Cenozoic. They suggest (p. 145) that the current high sensitivity 'is likely to persist into the future as the oceans become more acidic and the bulk of the fossil-fuels inventory is transferred to the ocean and atmosphere'. They do not harbour much doubt concerning the value of the Paleocene–Eocene Thermal Maximum as a guide to current affairs: 'This event provides a useful analogy for the present-day scenario of warming from anthropogenic emissions.'

That high sensitivity is more bad news, for as Cohen reminds us (see end of previous section) we have not yet seen positive feedback from release of methane from hydrates. That could be crucial: 'Because the ocean methane hydrates comprise a large pool of potentially releasable carbon, they have the potential to have a strong, long-term impact on Earth's climate' (Archer, Buffet and Brovkin, 2008, p. 6). Where does that leave us? Professor David Archer captures the gist in the title of his (2008) book *The Long Thaw: How Humans Are Changing the Next 100 000 Years of Earth's Climate.*

It is perhaps a little tempting to cut a few corners in advocating a case for action based on a particular line of evidence, to favour one witness rather than another. Honourable lawyers, politicians and scientists should of course firmly bid that particular devil behind them. So it is not claimed in this book that we understand the significance of all the evidence we now have to hand that bears on the 55 Ma warming event. Nor is it claimed that the 55 Ma event was in the very strictest sense an 'analogue' for present-day activity: it is simply a natural event from which we may draw with some confidence important conclusions of practical significance to humankind. From reading the 55 Ma record in the rocks we can see what will happen if we continue

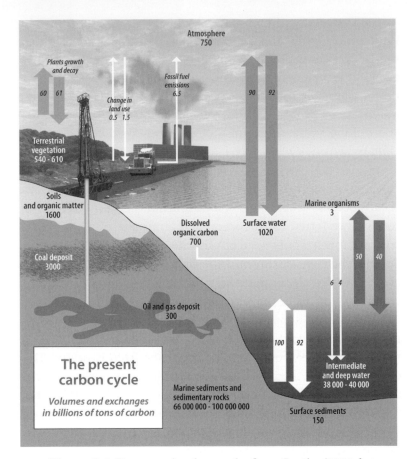

Figure 2.6 The annual carbon cycle, from Gautier (2008, her Figure 4.1). The figures shown for 'Change in land use' and 'Fossil fuel emissions' suggest that our influence on this cycle might not be great, but our cumulative effect over decades is notable (see text and compare with Figure 2.3).

to release a lot of carbon dioxide to the atmosphere in a short period of geological time.

The perspective in *Science* of Pagani and colleagues (2006), and the views of Goodwin and colleagues (2009) published in *Nature Geoscience*, are important parts of a fascinating and rapidly developing line of cross-disciplinary debate and research. This is the kind of research scientists really enjoy, with lots of enticing complications: in the language of science (see earlier quote from Harry Elderfield) the system being analysed is 'non-linear with feedbacks' – delightfully complicated. But the key point emphasised in this book is that all this research has a central theme from which it is hard to escape: a warning message quantified by Norris and Röhl and spelled out by Dickens at the end of the last century.

That message is simple enough. Releasing carbon at the rate we are now doing runs the obvious danger of causing major disturbance to the climate of the planet and to many of its other features. Carbon is crucial for life in general and any significant disruption in its cycle on Earth (Figure 2.6) is a real threat for *Homo sapiens*. Regional prediction of global sea-level rise following partial or complete melting of polar ice-sheets is not straightforward (Milne, 2008), but a significant overall impact on our worldwide community is beyond dispute (compare with the comment of David Jenkins in his 1997 memorandum, Chapter 1). And acidification of the oceans (Caldeira *et al.*, 2007) to the degree experienced at 55 Ma would present life on the planet with significant stresses from which *Homo sapiens* would not be exempt.

Messages take a while to filter through, and what is written in the rocks is no exception. A four-page letter to *Nature* by Norris and Röhl in 1999 was not going to bridge rapidly the Atlantic Divide in the oil industry publicly established at Kyoto in 1997. But we should have little doubt about the significance of the message itself.

By natural means the planet has already demonstrated 55 million years ago the damaging effects on life on Earth of releasing rapidly a large volume of carbon into the atmosphere. It is now plausible for the geological community at large, not least those in the oil industry, to join with the climatologists and conclude that if we continue to release carbon dioxide into the atmosphere at the present rate we shall, this century, experience among other effects significant acidification of the world's oceans and an overall global rise in sea level. Even at the lowest likely levels these changes will have a significant adverse effect on our species and at their upper likely levels would be disastrous for many of us. How will the oil industry react?

3

An Atlantic divide in Big Oil

3.1 A RIFT DEVELOPS

In considering climate change, geologists tend to look far back in time by the standards of human history, while continuing to use the present as the key to that past. The rest of the world tends to consider the record of climate over a relatively short period back in time, while concentrating on looking into the near future, using computers to help their speculation. Until the very last years of the twentieth century, the oil industry, with its strong geological heritage, tended to look back and reassure itself it was doing no harm. More than that, major oil companies were supporters of an organisation founded by some original 'climate-sceptics': the USA-based Global Climate Coalition (GCC). The now-defunct GCC was founded in 1989 by companies provoked by the formation of the Intergovernmental Panel on Climate Change (see Chapter 1) into opposing action to reduce emissions of greenhouse gases.

BP left the GCC in 1996, the first oil company to do so. Then, in April 1997, came Shell's (1996) Annual Report:

> There are still many uncertainties about the impact of increasing atmospheric concentrations of carbon dioxide on global climate. However, there is now sufficient scientific evidence to support taking prudent precautionary action.

Meanwhile, in January 1997, David Jenkins, then Director of Technology at BP, had sent his memorandum on climate change (Chapter 1) up to the managing directors. In this memorandum Jenkins made the business case for taking seriously the issue of anthropogenic influence on climate change and for making a public statement concerning BP's new position: '... whilst a warming world continues to be a political concern, the hydrocarbon industry should be prepared to respond in a constructive manner'. History should record that Shell had the priority in publication in breaking ranks with the rest of the oil industry. But the public front-runner was BP.

In May 1997, John Browne, Chief Executive Officer of BP, made the speech at Stanford University that changed forever the comfortable public consensus within the oil industry on climate change (Browne, 1997):

> ... over time we can move towards the elimination of emissions from our own operations and a substantial reduction in the emissions which come from the use of our products.

The aim to take responsibility for emissions from BP's own operations might be regarded simply as good contemporary practice, but Browne's second promise, to take similar responsibility for the far more significant and general act of taking fossil carbon from the ground and putting it into the atmosphere via the company's customers, was truly a gauntlet thrown down.

Why did BP make that move? As a former Chief Geologist of BP, Jenkins followed a long company tradition in being particularly open to ideas from the geological community at large (Lovell, 2008). The views of Jenkins and other senior staff within the company, rather than external pressure, led to the change in policy (Reinhardt, 2000). By the end of the century, younger members of staff were also influencing events. Records of attitudes of BP's international graduate recruits were kept by me from the mid-1990s (Lovell, personal files, available for study on request). These records show an initial regional variation in the recruits' views on anthropogenic climate change, with a 'sceptical alliance' between North American recruits and those from developing countries. Texans would reflect historical and contemporary attitudes in the oil industry in the USA, while Angolans might suggest that their country had other more immediately obvious challenges than any longer-term effects of climate change. Over the last decade this scepticism has faded to the point where BP's policy on climate change is now seen as an important positive factor by many recruits: this evolution is reflected strongly in the balance of opinion in formal debates on climate change during regular induction events involving recruits into oil exploration and production, the 'upstream' part of the oil industry.

So it is appropriate that a likeness of Browne, portraying him as a particularly happy latter-day hippy, complete with a helios wand, was published in *The Times* newspaper in London on 7 August 2000, as a comment on the greening and re-branding of BP. This cartoon illustrates an evolution in the public image of BP, reflected in the way new entrants regarded the organisation that they were joining. Earlier in the 1990s, Browne had warmly encouraged the new *Challenge* programme of graduate recruitment and development in BP (Lovell, 1998). The new recruits, attracted to BP by the scope of this programme, added to the internal pressure on environmental matters and advised on the re-branding.

Whatever the background reasons for the moves of BP and Shell in 1997, the changes in policy came at a sensitive time, given the imminence of the Kyoto Climate Summit. So the Atlantic Divide in the oil industry immediately assumed a major political significance. The scientific community took sides: '… there is no discernible human influence on global climate at this time' (Gerhard and Hanson, 2000, p. 466). This comment was published by the American Association of Petroleum Geologists (AAPG) and may be taken as reflecting a widespread view in the oil industry in North America at that time. Correspondence in the house magazine *AAPG Explorer* suggested that it was certainly a popular view for many in the AAPG.

The Chief Executive Officer and Chairman of ExxonMobil, Mr Lee Raymond (2000), took his view of the state of the science to a business conclusion:

> We agree that the potential for climate change caused by increases in carbon dioxide and other greenhouse gases may pose a legitimate long-term risk. However, we do not now have a sufficient scientific understanding of climate change to make reasonable predictions and/or justify drastic measures.

The more implacable opponents of the approach taken by ExxonMobil would quarrel with Raymond's view of the state of the science, believing that the company's position was politically inspired. Even if that were true, it would not be a helpful attitude. There is an argument that says we cannot run computer models of future climate with overweening confidence because we do not yet properly understand the relationship at Earth's surface between the huge input of heat from the Sun, the effects of the relatively small input of heat from Earth's interior and the parts per million of carbon dioxide in Earth's atmosphere (see discussion in Chapter 2). Might this lack of scientific understanding justify delay in controlling anthropogenic release of greenhouse gases?

That question was the essence of a debate that involved science, business and politics and characterised the Atlantic Divide in the oil industry at the beginning of this century. On one remarkable occasion that debate became direct and public. On 26 March 2003 senior representatives of BP and ExxonMobil took their positions on stage in the Geological Society lecture theatre at Burlington House, Piccadilly, London. What happened that evening gives the most useful and lasting insights into the topic of this book and is here considered in some detail.

3.2   BP AND EXXONMOBIL DEBATE: A CLOSING OF THE GAP?

The debate took place at an international conference on *Coping with Climate Change*. This meeting was convened by the Geological Society's Petroleum Group, a well-stirred mixture of oil folk, government scientists and academics: my fellow-convenors were Dr Lidia Lonergan from Imperial College and Dr Nick Riley from the British Geological Survey. The purpose of the conference was to review the main lines of evidence for and against man's influence on climate. Nothing very new there; there had already been plenty of scientific and political debate on this topic, although not on the scale that was to follow in the next few years. What was different in the 2003 meeting was the emphasis on two things. One was the restraint imposed by the geological record on modelling and speculation about future climate change. The other was a practical recognition that climate has changed often in the past, on some occasions rapidly, and will change in the future whatever we do about our own contribution to greenhouse gases. Hence the emphasis was on coping with climate change as well as discussing its causes and effects.

On the first day of the meeting, attention was on reading the geological record of climate change. Recent advances in understanding how to read the story of climate change from deep-sea sediments and ice-sheets were given due prominence. On the second day, the impacts of those climate changes were considered; impacts on land, in the ocean and at the coast. By the third day we were concentrating on the practical question; when we are looking for solutions to the problems caused by climate change, can geology help? On this third day the focus was on matters where rocks are undeniably in centre stage, at least in Act One of any drama with the title *Coping with Climate Change*. There was emphasis on the so-called technological fix of burying nuclear waste or excess carbon, as well as on mineral sequestration of carbon dioxide.

In the midst of this sequence, on the second evening, an ancient tradition of the Geological Society was sustained: a public debate took place on the causes of climate change and how to cope with its consequences. This discussion was regarded by the convenors as much more than a nostalgic excursion into the style of meeting held in the nineteenth century: we were motivated by a belief that the modern oil industry lies at the heart of the application of science in response to the challenges of climate change.

The debate was between Dr Frank Sprow (Figure 3.1), Vice President, ExxonMobil and Mr Greg Coleman (Figure 3.2), Vice President, BP, the senior representatives on environmental matters of two major organisations publicly separated by an ocean of difference on this matter since 1997. This difference is for many people epitomised by the support of BP for the 1997 Kyoto Protocol, compared with the opposition of Exxon and Mobil (Leggett, 1999). In this opposition the two USA-based companies, since then combined as ExxonMobil, anticipated the policy on Kyoto to be adopted by George W. Bush on becoming President in 2001. In his comments in the debate, Sprow explained the ExxonMobil position:

> Kyoto … is a lot of fuss about something that really does not make much difference.

Figure 3.1 Dr Frank Sprow, Vice President of ExxonMobil, presenting his case at the debate between BP and ExxonMobil at the Geological Society in London on 26 March 2003. Photograph by Ted Nield.

Setting the specific arguments about Kyoto to one side, it was clear from the opening remarks that BP and ExxonMobil shared a deal of important ground. The extended quotations that follow are taken from the two opening speeches and are here juxtaposed for comparison:

BP (Greg Coleman) on the company's priorities

We think about what contribution we can make as a company and as an industry in terms of security of supply, whether it is from crude oil, natural gas or any other form of industry. We believe that one of our responsibilities is to ensure that our customers have a reliable supply of energy when they need it, regardless of what they are expecting from us in terms of the price – that tends not to be a consideration. We do believe that

Figure 3.2 Mr Greg Coleman, photographed by Ted Nield at the BP–ExxonMobil debate in London on 26 March 2003.

there is a direct link – and there is much empirical evidence to support this – between economic growth and social growth. Lastly, we do believe that is it very important that, as a company and as a society, we preserve the natural environment. We think that we have a role in all three of those and I will say a bit about what we are doing in those three areas …

ExxonMobil (Frank Sprow) on the company's priorities
I thought it would be worthwhile to take a moment and state our position on the subject of climate change. I think that, from time to time, this is not terribly well understood and I take all of the blame for that. I think that means we are not doing a very good job of articulating it, but here is another try. I think that you will note some common themes with Greg's comments. We feel that certainly for a long time – we would say until mid-century – fossil fuels are very likely to remain the

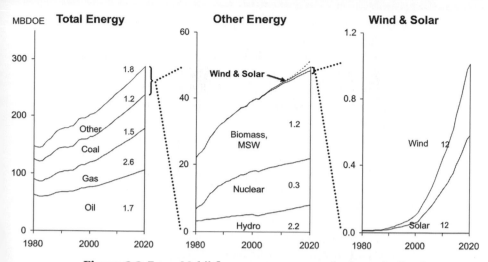

Figure 3.3 ExxonMobil figure on energy supply shown by Frank Sprow at the March 2003 debate. The vertical axis shows demand as 'millions of barrels a day of oil equivalent' (MBDOE). 'MSW' is municipal solid waste (garbage). Even increasing the growth rate of energy from wind and solar by an order of magnitude from the 20% a year assumed here makes little difference to Sprow's main conclusion: for at least the medium term we are bound to rely heavily on fossil fuels.

dominant source of energy [the slide shown by Sprow at this point is here Figure 3.3].

The world needs affordable energy to drive economic growth and human progress and to a large extent, that is going to take fossil fuels. Thank goodness for geologists, who keep finding a fair amount of fossil fuels. I should emphasise, however, that meeting this demand will require our best efforts. One thing that is often forgotten in this overall discussion is that is takes a good deal of work to keep finding fossil fuels faster than you are using them. Our projections would show – and I do not think these differ very much from anyone else's – that about half of the oil and gas that will be needed in ten years is not yet in production. The other thing that might be said, and perhaps today's international situation makes this really clear, is that many parts of the world where oil and gas reserves remain are not particularly friendly places to work. The bullets are flying in many of them from time to time. So just the fact that the oil is in the ground does not mean that it can be extracted in a way that will put in at the fuel pump or in the heating oil tank.

The second point is that by our analysis – and I do not think this differs from most other analyses – renewables are non-economic. In niche markets, you can get cases where solar cells make sense and where windmills provide an attractive source of energy but by the by, and I will come back and talk about energy forecasts in a bit, they are non-economic at this point. A significant role for renewables is only possible with substantial government incentives and mandates, which basically means that the taxpayer is paying the bill instead of the customer. A societal question is 'Where is it best for governments and others to spend their money?'

The third point is that whilst we may pick about the climate science and talk about the uncertainties that are there, which we have done and do and probably will continue to do as long as those uncertainties remain, the important thing to us is that the risk of adverse climate change is high enough to take strong and effective action. That is our conclusion on dealing with the risk. We also note that the actions taken to mitigate climate risk must take into account their impact on economic growth and impact on societal values as well …

BP on emissions

… Just to articulate what our new plans are, we met our previous target and now we are in a phase where we think we have captured the low-hanging fruit. Consequently, we are setting on a track of providing the world with solutions to allow us to stabilise the level of carbon dioxide in the atmosphere. We are not in a position to say what that level might be but it cannot continue increasing at the rate it has been for the last fifty years. There needs to be some will on the part of governments and society at large to reach stabilisation, and we believe that is an important decision yet to be made …

ExxonMobil on emissions

… We also believe – and Greg mentioned this one as well – that if we are going to improve something, if we are going to reduce greenhouse gas emissions in a big way, we need to measure them and we need to have everyone reporting them. This includes generators and consumers and we need to account for all the greenhouse gas emissions. We think that should be done on a mandatory basis through government reporting and we also think that it should be done on an equity basis. There I mean that if you are an investor in a plant that is operated by somebody else, you should carry the burden for your investment in terms of the emissions that it generates. So we end up counting everything, as opposed to relying totally on operators to generate that information …

BP on carbon capture and storage

… The bottom right of this slide [here Figure 3.4] shows our favourite subject and one that I know is close to the hearts of all geologists. This is

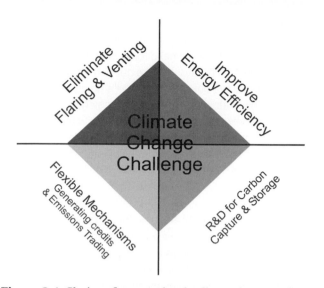

Figure 3.4 Choice of approaches leading to lower carbon output, shown by Greg Coleman of BP at 2003 debate. Flaring and venting in BP's own operations are to be distinguished from the other three categories, which prospectively involve BP in taking responsibility for the use of its products by its customers.

the use of carbon capture and storage. We do not advocate storage in oceans, although many of the people in the science world are talking about storing significant amounts of carbon dioxide in the deep oceans. We do not believe that is something that we, as a company, can make a contribution to and, given the lack of understanding of the whole carbon cycle, we think that it is an area that needs to be studied more before we decide to do anything. I will say a bit more about carbon storage and then I will say something about the last area that we will focus on in a moment: flexible mechanisms …

… Our internal calibration is that we are talking in single digits dollars per tonne to store carbon dioxide in a reservoir. When we put in the cost of capturing carbon dioxide, the cost can go up significantly if you are capturing the carbon dioxide from dispersed sources. If it is a concentrated source, it is still fairly attractive compared to some of the alternatives …

ExxonMobil on carbon capture and storage

… It does not need any additional boost at this meeting to emphasise that means to sequester carbon dioxide emissions from central sources are critical. Whilst it is very important to do that work in ways that will ensure that we do not threaten the environment more than we help it, the reality is that we are going to have to use a lot of coal if we are going to have the energy we need for at least the next 25 or 50 years. I say that without bias because we are out of the coal business and I am delighted that we are out of it. But there is going to be a great deal of coal used at generating stations if we are going to have to power that we need. The only way that can be consistent with a carbon-constrained future is by the capture and sequestration of those CO_2 emissions from those large generating stations. The geology here is likely to be critical.

BP on solar energy

Just to conclude, in terms of all of the things that we might be doing to reduce our carbon intensity, we are continuing to invest at a significant rate in our renewals business, particularly focusing on solar energy. I think most people know that it is not really cost competitive today unless there is an incentive provided by local governments to actually create the local demand for the product and to allow people the capacity to develop it. We are investing in new technology and we are continuing to rationalise and increase the efficiency of the way we produce our solar energy or solar panels, but we have some way to go before it will become cost competitive with hydrocarbons …

ExxonMobil on wind and solar energy

… So here is wind and solar up top and our best projection is that little green band up there [in 'Other Energy' in Figure 3.3]. Actually, it is pretty bullish so we have gone through and looked at the technologies, the rate of technological improvement and we have made some fairly bold assumptions about how technology and applications can proceed. But let us forget about the little green slice and assume that wind power and solar power can grow by 20% per year. Then you get this chart over on the right with wind and solar, but even with strong growth you can see that it is unlikely with anything like today's technology assumptions that they are going to play a very large role.

BP on governments, partners and customers

… Lastly, just to talk about partnerships, there are a number of different players that need to be involved in making something actually feasible and practical for the world so that we do not have to slow down or stop economic growth. We need governments to be involved and the UK government is taking a lead in having a UK emissions trading system. It actually subsidised it, if you will, by contributing money to get the system going. I think that learning by doing continues to be something that we believe is valuable.

The automobile manufacturers and general industry are making significant progress. Both we and Exxon, and most of the companies in our industry, are working actively with the automobile manufacturers on anything from engine technology to the simple process of burning hydrogen. We are continuing to learn things about what that efficiency is and how hydrogen actually burns, which is something I thought we had figured out a long time ago. As I mentioned earlier, we have significant programmes at Princeton University [USA], Tsinguha University [China], Cambridge University [UK] and a number of other universities in the United States, looking at both the science of climate change and the technology programmes that can help significantly reduce the cost of dealing with greenhouse gases.

Customers, at the end of the day, have a most significant part to play here because they are creating the emissions through their use of our products. We think it is a factor of ten times what we are generating in our own internal operations. Getting customers to recognise that there is a cost in the way they use energy is important, whilst also recognising at the same time that customers are not prepared to pay a premium to have a cleaner and greener fuel. At least they have not demonstrated that they are prepared to pay a premium. So we have some work to do there. Last, but not least, we have found NGOs [non-governmental organisations] to be very helpful and constructive in coming up with solutions and in challenging us to think creatively and to move into action.

ExxonMobil on university partnership

… One thing that we are doing – something that I am particularly proud of because I spent a lot of time helping to put it together – is a very large programme called The Global Climate and Energy Project [GCEP]. If you want to find out about it, go to www.gcep.stanford.edu. It is a programme that is managed by Stanford University but will involve the efforts of a number of other universities in Europe, the US and in Asia. It is not on climate science but rather on innovative technologies to reduce climate risk. We are just beginning to put the programmes in place now but I am really excited by what is going on. What is unique about this programme is that it is one that hardwires companies back into the process, not in terms of dictating what they think are technology solutions but rather hardwired in terms of using the output.

The companies that are part of this programme include ExxonMobil at the $100 million level, Schlumberger, which got by with a mere $25 million – I do not know how we let that one slip by – General Electric at $50 million and Toyota at $50 million. That is a lot of money; that is well over $200 million. Now it is true that there is no university that has the amount of money they want but I think this will make a big difference. We are going to establish an advisory board comprised of non-industrial participants to guide the work, to provide input to Stanford and the sponsors on the most attractive areas of endeavour and to establish the quality of the work that is going on …

Author's comment

Following the two opening statements, the session then moved to questions and answers, led from the chair by Mark Moody-Stuart, President of the Geological Society. This section sets out the points of agreement and disagreement. It covers matters of lasting significance that are still largely unresolved. These matters include renewable energy, emissions trading, market mechanisms, legal constraints, government regulatory frameworks, the Intergovernmental Panel on Climate Change, taxes, and hydrogen as an energy carrier.

The content and the tone of the discussion reflect its unscripted nature and give it particular value. The discussion gives an unusual insight to attitudes in the oil industry on a range of significant issues connected with climate change. It is here quoted in full so the reader can get a proper view of the strands woven through it:

> **Mark Moody-Stuart**   Thank you very much indeed Greg and Frank. I thought those were both very interesting overviews of the situation as viewed from different corporations. I do not know about you, but I was ticking off the points of agreement and the possible areas of divergence. To me, it seems that you are both saying that there is an issue that needs to be addressed. You are both keen on controlling your own emissions, reporting your own emissions and ensuring that they are accurately recorded. You both acknowledged that the major problem is not actually in your own emissions but in what customers do with our energy – it is the 'about 87%', as you said. So I think those are all points of agreement.
>
> On points of difference, clearly there is a point a difference, which you can see physically, on renewable energy. One invests in it and the other does not at the moment. The one who does not says that it does not

make any money and I suspect the one who does will agree that it does not make much money. That is my experience.

Frank Sprow   We like BP spending money on things that do not make much money.

Mark Moody-Stuart   And then I detected a bit of difference on the role of government in channelling this. On the one hand, Greg referred to necessary government frameworks and Frank, rather more pejoratively, referred to the fact that renewables would only work if you subsidised them. You are both looking at longer-term sequestration and technological activity into new areas, for example hydrogen and so on. I will leave it now to you to see if I missed anything.

Greg Coleman   The only other thing I would point out is that we are much more interested in seeing emissions trading systems developing and I do not see Exxon as being as active in that area. I do not know that you are not active, but I do not see much involvement from Exxon in emissions trading or working with some of the regulators that will be regulating some of these emissions trading systems, which we think allows us to be able to optimise across a global portfolio more easily. One of the reasons why we believe that it is appropriate is because we believe it is the only way to ensure that the capital is deployed most effectively. There is going to be a lot of capital involved in dealing with this and we feel that it is very important to make it most cost efficient by putting it in the most effective place. Another difference, and Frank pointed it out, is the way that we manage our different companies. We tend to have more of a target setting/look forward approach. I am not sure how I describe what you do, Frank, but it might be interesting to hear your views on how you have managed to achieve the energy efficiency that your company has gained.

Mark Moody-Stuart   Would you like to say something about Exxon's approach to trading and global market mechanisms or national market mechanisms?

Frank Sprow   We do a lot of trading and I think we are good at it. In areas where trading is required, as is increasingly the case in the UK – we are talking about carbon emissions trading – I think we will do it well. Even though we do not have the head start that BP has in this area, I have enough confidence in our people to think that we will be very adept carbon traders if that is required. For us, I think the matter of emissions trading has, for better or worse, got mixed up with Kyoto, which we think is a very bad idea. In parts of the world where Kyoto is ratified, we will be fully in conformance with it and we will optimise our business under those constraints.

But I do think that it has been hard for us to do much to talk about emissions trading without getting over the wall on Kyoto. I would like to see those two things separated. Emissions trading can play a role in optimising one's own economics and driving the right kinds of things in one's operations. However, I would say that at the core that emissions trading, in our view, is sort of ensuring that we are harvesting the low-hanging fruit on the tree effectively. In our view, and I suspect in BP's as well, to make a real dent in the issue of reducing climate risk, it is the majority of the fruit high on the tree that will need to be harvested. That is going to take advanced technology and probably many other approaches. I guess we view emissions trading as having an excessive overlap with Kyoto and also being a bit distracting from getting on with the main job of doing the science and technology needed to make a real difference. I am just giving you an idea on this, but that is how we approach emissions trading.

Mark Moody-Stuart   Given the fact that Exxon is reducing its own emissions, reporting on them and would probably be, in own emissions, more than capable of dealing with Kyoto, what is it that you dislike so much about Kyoto? Why is it not just a harmless effort to start doing something?

Frank Sprow   I will make a comment about US law first and I hate to start with lawyers – that is a horrible place to start – but that is what came to mind immediately. I think Europe tends to be a 'best efforts' kind of society. If you say that you are going to cut emissions by 5% and you put in it your report and ultimately you do not succeed, either as a company or a country, in general the regulators will say 'We know you tried hard but it did not work out so better luck next time.' In the US legal system, if you put something down as a commitment, or if the country puts something that is obviously important to us down as a commitment, and you do not make that commitment for whatever reasons, we will have people lined up in the courthouse to sue us for not meeting that commitment. Now you could say that is a good recognition of the US legal system, or it is appropriately paranoid on my part, but that is the reality of life in the US.

The other thing that I would say about Kyoto, now that you have got me going, is that it is a lot of fuss about something that really does not make much difference. If the international community is going to work on ways to mitigate climate risk, it is going to have to focus on science and technology. It is going to have to focus on developing countries and it is going to have to focus on a market basket of activities that have the potential to make a real difference in reducing carbon emissions to a level much lower than the current level.

To have countries focused on whether it is 7% or 5% or 11% is, to us, basically an international distraction. While it may get the ball rolling, is it rolling in the right direction and doing the right things? We are sometimes accused of being US-centric but I hope that is not right because we do business in 200 countries so I certainly do not think that I am US-centric. We are a little fearful of legal obligations – that is undoubtedly correct – and we also think that it is somewhat distracting from an international policy point of view. That is my anti-Kyoto speech for the day, perhaps.

Mark Moody-Stuart   I am sure we will have some response later. Greg, you mentioned government frameworks, which I believe are absolutely essential. Some of these things are not going to be delivered purely by the market. Could you say something about BP's approach to that – not necessarily carbon taxation but perhaps regulatory frameworks?

Greg Coleman   First, there is some immediate experience that is under way in the United States right now, with the regulation of sulphur dioxide and nitrous oxide emissions. The amount of capital that is being invested by industry in the US to deal with these issues is billions of dollars. Unfortunately, as Frank has pointed out, the regulatory system in the US is pretty onerous but it has also focused all of the effort on places where there may not actually be the best opportunities. We believe that there is a critical need for an international framework to ensure that the best opportunities are the ones that are pursued. There are not that many international frameworks, especially in the financial systems. The WTO [World Trade Organization] has been trying to get reduced tariffs in trading and it is proving to be very difficult. But the alternative is to have every country fighting their own battle and we think that does not lead to a solution that is going to be quick enough or able to deal with all of the social impacts that we will have. We also believe that incentives are the right approach, as opposed to penalties. We do know that governments that use penalties, called taxes, to make things happen have a big need for money and they will do what they think is right with that money once it comes into their account.

We would rather that the market players, not just companies but all of the players including users of capital, have a chance to play off against each other. We think that it is important that the NGOs, the regulators and the communities that are affected play a part in the solution, as opposed to having it imposed. So although we would like governments to provide a framework, we would also like them to engage all of players in the development of those frameworks.

Frank Sprow   I might make a comment on that, if I could. I am going to say something about Kyoto again, and I apologise for that. In terms of the international or the governmental process, I would just like to make a couple of points. We do not think Kyoto makes sense as an agreement. We are definitely not against – and very much support – the Framework Convention and the idea that the only way this risk is going to be mitigated is through a successful international process. So I really want to differentiate those two things.

Whilst we sometimes quibble with the way some of the IPCC results are quoted in a political context, we very much support the IPCC process. We are involved in it and I can assure you that when the IPCC reports come out, we look them over. I, to my limited ability, look them over and I find that I am in agreement with 99.9% of the material in the individual scientific contributions. So that is not the question.

The role of government will be particularly important, as Greg points out, in trying to figure out how we are going to make things happen that are good things in reducing climate risk when it is not obvious who is going to pay for them. Sequestration is one of the best examples of that because, as I mentioned earlier, we may well need to be able sequester CO_2 emissions from large coal-fired central generation stations, assuming it can be done in an environmentally sound way. Now that sounds really good but the coal companies, and to a large extent the utility companies, do not make much money.

So you have got what could well be the biggest need for action coupled with those with the thinnest pockets. How you pull those two things together to provide the incentives and the required actions clearly has to involve governments, one way or the other. I think that is just a classic example, to make your point Sir Mark, that governments have to be part of the solution.

Mark Moody-Stuart   You could see, for example, a climbing requirement, maybe coupled with some kind of tax rebate, for sequestration. For example, something like starting with a small amount and then climbing.

Frank Sprow   I am not allowed, as a good American, to use the words 'tax' and 'government' in the same sentence but we understand the point you are making!

Mark Moody-Stuart   I was talking about tax rebates.

Frank Sprow   Oh, that is all right.

Mark Moody-Stuart   Earlier, someone remarked on this question of tax neutrality. Most of us in the industry, when we hear tax neutrality, we do not believe it.

Frank Sprow   How that is dealt with is going to be an amazing question.

Mark Moody-Stuart   What about hydrogen? I get the impression, and this may be wrong, that Exxon and indeed the United States is banking on hydrogen more than BP is.

Frank Sprow   Greg and I discussed hydrogen over dinner last night. Our view is that hydrogen is not an energy source, it is an energy carrier. When we talk about hydrogen, particularly in a renewables context, we are really trying to take water and, at great cost and expense, split it up and produce hydrogen, and then put that hydrogen into a devilishly complex system, like a fuel cell, to make probably not quite as much energy as we put in and water. If you wanted to be a little cynical, you could look at this as a very costly water-recycling system.

I am going to be more open minded than that because clearly one has the potential, using hydrogen as a carrier, of having the vehicle and the fuel free of pollution in all of its normal concepts. This is such an intriguing target that you have to, as people with technological confidence, be interested in pursuing it. How can we produce low-cost hydrogen, either from natural gas or perhaps ultimately from renewables? How can we deal with the safety issues in hydrogen, which are very formidable? They are too easily dismissed but the idea of rather clumsy consumers – I do not mean anyone in this room – going to an unstaffed store and putting a high-pressure hydrogen line on their vehicle scares me a lot because I am in the safety, health and environment business.

I am not saying that with anything other than a lot of conviction. Is there the potential to deliver that hydrogen in a solid form, for example a hydrate or some other structure, which could be done at much lower risk? I certainly hope so. Are we enthusiastic enough about hydrogen that we are involved, as is Shell to a greater extent than we are, in some government demonstrations of hydrogen? Yes. Do we have a lot of concerns? Yes. Can I speculate now on whether hydrogen is likely to be a winner or a loser? Quite frankly, I do not know but we are spending money to find out.

Mark Moody-Stuart   Okay, I am about to hand over to you but before I do that, could I ask each of you a question? On the IPCC scenarios, you have various outcomes. If you take the low scenarios, which result in significant climate change but not as scary as the top ones, is the lower end of that doable or are we stuck on the higher end? Do you see a way in which the energy industry, which is responsible for most of this stuff, could actually deliver a world whereby carbon dioxide stabilises at 550 or 600 parts per million [ppm] greenhouse-gas-equivalent, or do you just say that it is in the laps of the gods?

Frank Sprow   I do not know which of those scenarios are most likely or if in fact the actual scenario could be lower or higher than that range. I would see the band of uncertainty as being quite significant. I would say that when we pulled together the GCEP programme that I mentioned, we all did some back of the envelope work where we looked at different scenarios and processes. We have belief in science and technology and time to reduce climate risk, and believe there is nothing terribly frightening out there if we get on with it and do the right things. I think that if we waste a lot of time, that is the trap for us.

Greg Coleman   Our company's position is that we believe that the planet needs to achieve stabilisation and that it can be done. It is going to require efforts on all fronts and, much as Frank has said, we have to move the whole piece to achieve that end. The work that the Princeton scientists have done does say that there is still quite a bit of uncertainty and it is not a one-way direction either. The CO_2 can go up into the atmosphere and it can be reabsorbed back into the oceans, depending on the ocean cycles and things like that.

But I think we are optimistic because we have seen how much we have achieved in such a short period of time. At the same time, we need everybody to be moving in locked step pretty quickly. The longer these debates go on where people seem to be arguing about something that has a minor impact, the less attention we can put on the fundamentals.

Mark Moody-Stuart   Well, you have seen a fairly cosy interaction between us energy folk up here and now over to you.

A number of comments from the audience challenged the cosiness identified by the chairman:

Participant Eight (unidentified)   I have got a challenge for you. My impression was that the investment that your companies are making in this climate change issue are more marketing gloss to keep the NGOs off your backs whilst you go about making money out of oil and gas. I suspect that your core competence is in finding and producing oil and gas and that the companies in the new arena are not going to [inaudible] BP and Shell.

Mark Moody-Stuart   Well, there is a challenge.

Participant Nine   I had two thoughts also about renewables. One was that when Frank says that they are uneconomical, I assume that he actually means that they are profitable because they are only uneconomical if you do not cost the externalities involved in other [inaudible] production. To say that they are uneconomical is an aspect of bookkeeping; it is an aspect of under-pricing the externalities.

Participant Ten   … in America, and in many countries in Europe you can go into a filling station and fill up with hydrogen quite safely. I think the second point is to do with externalities. You have mentioned the current conflict in Iraq. The US government spends hundreds of millions of dollars protecting oil supplies from the Middle East and both the US and UK government gain security of oil, one of their top foreign policy objectives. It already costs the taxpayer very large sums of money to protect oil supplies, which of course the oil industry gets free of charge.

That also goes with the cost of climate impact and so on – it is not paying for that.

Professor Miriam Kastner, Scripps Institution of Oceanography   It is interesting that despite the fact that in several recent speeches President Bush has mentioned hydrogen as one of the important future energy sources, BP seems to have a more positive attitude about it than ExxonMobil.

Frank Sprow   That might show that ExxonMobil and President Bush are not as hardwired as some might assume.

Dr Eric Wolff, British Antarctic Survey   I heard from ExxonMobil that they believe that most of the science about climate change has been presented but there are still some uncertainties. It seems to me that those uncertainties get propagated and amplified if you go upwards to the chief executive and chairman of Exxon, and then eventually until you get to the President of the United States and so on. They then provide an excuse for not treating the problem very urgently.

You said that we should unite around solving the uncertainties so I would turn that around and ask what do Exxon see as the uncertainties that are stopping us from acting urgently? We know what we as scientists see as the urgent uncertainties, but we are not sure that they are the ones that are stopping people from believing the stories that we are telling. So we need to be told, as climate change scientists, what are the things that are seen out there as being uncertain?

Bryan Lovell, University of Cambridge   I just wanted to say that I provoked this debate originally in the Petroleum Group, hoping that there would be a prize fight when we came to the public spectacle. In fact, we have had the boxers embracing at the weigh-in at the restaurant yesterday evening before they even climbed into the ring here today. But I think we may have had something rather more exciting than a fight – we may have had agreement. If I may, Mark, I would just like to say why I thought there was a qualitative difference between BP and ExxonMobil that led us to this afternoon's discussion. I think that we have reached the overwhelming scientific consensus in this room, notably expressed by Hans Ziock [Los Alamos National Laboratory, New Mexico] at the climatic moment when he said we had not got any time left, that we think as a group that there is a big problem and that urgent action is required. I think there may be others here who share with me the belief that if your companies do not do something about this, then we really do not have much hope. I do not see anybody else besides the oil industry actually doing anything about it on the sort of scale required. So, can we have a statement of intent from ExxonMobil that they will emulate BP in taking responsibility for the carbon dioxide emissions resulting from the use of their products? Then we can start doing something as an industry about this problem. As long as there is not a clear recognition that the responsibility of the oil companies is for the use of their products, I do not think we have much hope. I don't give that much credit to people simply taking responsibility for emissions in their own operations. I run one of Britain's smallest companies and I certainly take responsibility for ours and I would hope anybody running an organisation would do the same.

Sprow's immediate response was, to many of us present, both a surprise and a big step forward by ExxonMobil:

Frank Sprow   Absolutely. Let me start with Bryan's point and then I am just going to choose three others here for at least an initial response. I think we very much do take responsibility for the use of our products. I think the programmes that we have undertaken with automobile manufacturers and other device manufactures are responsive to that. I think we were one of the first companies to point out that the lion's share of fossil-fuel emissions came from the use of products. That is not to mean that we should not do our best to reduce our emissions in our plants but it really is in product use that we will make a difference.

I think we were the first of the oil companies to engage with automobile manufactures in fuel cell research. We did that back in 1990 or thereabouts when we set programmes in place with Toyota and others. So we are definitely committed to working on improvements in the product consumption sector and we will definitely take your pledge, Bryan.

In his summary of the evening's debate and discussion, Moody-Stuart returned to that point. His closing remarks will stand not just over the few years since 2003, but for some time to come:

Mark Moody-Stuart   Okay, could I just say something about Bryan's challenge because I think it is very important? Given the position in which most of the major oil companies find themselves in relation to the work done by the IPCC, which is that it is a sound piece of science and it has a range of outcomes with error bars on them, it is a certainty that we quite often invest on. I think it would be a great help in addressing the consumer side if, in some way, we could collectively say that we acknowledge this work and we need to come out at the lower-end scenario, although we do not exactly know how we will do it. I do not think that we energy companies believe the person who said that we need to stop where we are now, and nor does the IPCC, I think. The earliest you could flatten out that is somewhere around 550-ish parts per million [ppm]. I think it would be constructive if collectively we could actually say that and say that we think, given that work and its uncertainties, that we need to level out somewhere around 550 ppm. There would immediately be a bunch of people who would ask why not 500 ppm or 450 ppm or whatever but frankly I do not care what the number is. I think a commitment would have an impact on consumers. The second thing I think we need to be quite clear about is something that I encourage business people to acknowledge – and this was raised at the back – which is that we do need these government regulatory frameworks within which the market can operate and deliver solutions, including flexible mechanisms. The person at the back said these things will not happen unless there is some channelling of the market and that has to be in transportation, in housing, in renewables and in fuel cells. Some kind of framework guiding the market and if we in major corporations could support that, it would help the consumers to support their governments in setting these things.

I mean we cannot sell things to our consumers that they do not want and governments, as Gordon Brown in this country found out, cannot do certain things to consumers because they rebel if they think that you have gone too far. Our consumers and governments' voters are the same people and that is where we need these partnerships to let us set up frameworks that allow us to work and deliver solutions within that. That is the second thing. So first, a collectively commitment to a target of some sort – that will take us a while, second, an acknowledgement of the role of government frameworks and third, this hugely important thing about the developing world.

All of the growth of transportation and so on is in the developing world. None of us are planning growth of fuel in the developed world, we are acknowledging that it will be flat or going down and how do we actually deliver those solutions to those folk? The opposition in Zimbabwe may be extremely enlightened and may be able to persuade people to go down a different track but I would bet that there is hardly a government in the developing countries that is capable of doing that. Developing consumers want what we have got so we have to demonstrate in the developed world that actually you can get the same utility and hugely increased levels of efficiency in our society, so as they go up the curve they do in fact leapfrog us. That is the third thing I think we should say.

Can I just draw your attention to the fact that tomorrow morning, David Jenkins is going to be giving a little bit of a summary and at the end of the day tomorrow, after a lot of very interesting talks, John Lawton, the chief executive of NERC [Natural Environment Research Council] is going to give us some reflections on the meeting as a whole. Bryan, is there anything more I should say? Thank you very much. On your behalf, let me thank our two debaters.

Come the end of the debate, the organisers were more than usually relieved and delighted to have had the three main participants on the platform and a full house to react to what they said. The habitual international travel of oil folk had just been greatly interrupted by far more significant activity in the Middle East: the second Gulf War had just started. On that same Wednesday evening Mr Richard Hardman, lifetime man of oil and former President of the Geological Society, was appearing on British television (BBC2) in the serious setting of *The Money Programme*. He argued with remorseless and elegant clarity, over maps spread out in the Lower Library at Burlington House (in a talk recorded before the day the BP–ExxonMobil debate took place there), that the war in Iraq was about oil. More than that, he suggested, the events in Iraq in 2003 might in the course of history come to be seen as one in a series of such conflicts, as demand for oil begins to exceed supply in the early twenty-first century. At the least, one would have to agree with Hardman that this was yet another conflict over control of resources in an uncertain world. But was the fighting over oil that should never be used anyway (Lovell, 2003)?

John (now Sir John) Lawton, then Chief Executive of NERC would say yes. He spelled out an intense view of the debate in the meeting's concluding address the following day. Lawton defined a spectrum of opinion, running from the damned, who denied the reality of climate change caused by man, to the righteous. The greater part of the audience of scientists before him was claimed by Lawton as righteous. The oil companies were consigned to the darker end of the spectrum, with BP slightly nearer to the light of salvation than ExxonMobil: 'At least with tobacco you can choose whether to smoke it.' Hardman had suggested that we were entering a period of conflict over finite supplies of oil; here was Lawton saying that much of the oil that remained on Earth should never be produced because of the damage its use would inflict on the planet.

Everyone who sat through the three days of *Coping with Climate Change* could feel the seriousness with which, in all their diversity, the scientists put their case. Equally obvious was the determination of these scientists to mitigate further damage caused by the atmospheric dumping of fossil carbon. But what of the oil companies, who are so much a part of the problem they must be an essential part of any solution? Could they agree on a common course of action? Only, it seems, if governments take the lead in setting a regulatory framework, as proposed by Moody-Stuart.

David Jenkins, former Technical Director of BP, in his review of the debate the following morning, proposed a government-led solution that tackled the carbon issue directly by offering incentives instead of penalties:

> The glaring gap in policy is any serious attempt to change consumer behaviour. Consumers change only when they want to or it is made demonstrably worth their while. To date there is no political will to effect any such change. In theory it could be very easy to do. In practice it is probably politically impossible. Incentivising conservation requires a completely different approach to that which penalises consumption. At present a focus on the latter automatically leads to revenue raising instruments. An objective function focussed on reducing greenhouse gas emissions requires relinquishing a mind set that views energy consumption as a source of tax revenue. Consumers would be set targets for reduction and tax penalties would arise only if the target was exceeded. Rebates would be available for reductions greater than the target. There is no sign that governments wish to adopt this approach. Were they to do so dramatic reductions in emissions could be achieved rapidly. Without it they won't happen.

Jenkins also repeated an earlier assertion (Jenkins, 2001) almost beyond dispute by any in his attentive audience, certainly by the oil folk:

> Climate change is a fact of life. The climate has always changed and always will. There is absolutely nothing humans can do to stop it. Stability is definitely not an option.

A formal announcement by ExxonMobil on 4 February 2004 followed the debate and covered much the same ground as Sprow had done in London the previous March. Meanwhile the Geological Society's house journal *Geoscientist* was claiming that the debate represented a significant closing in the Atlantic Divide in the oil industry. When I tried to convey this view to my fellow members of the American Association of Petroleum Geologists (AAPG) by submitting a manuscript to the *AAPG Bulletin* I got a cool response from the reviewers and a detailed letter of comments from a colleague in ExxonMobil. I had overdone the politics and underdone the science, manifesting to my American colleagues as an environmental zealot rather than a detached scientist. Subsequently I calmed down and the *Bulletin* published an exchange of views on the science between me and Dr Lee Gerhard of the Kansas Geological Survey and the University of Kansas (Gerhard, 2006; Lovell, 2006).

The tone of the London debate was important and the transcript of what was said there, much quoted in the preceding section of this book, has been audited and approved by the participants. The seriousness of the commitments made in Burlington House that evening remains to be tested by events, but commitment there was. Indeed, protestations of environmental virtue concerning climate change by oil companies have become a commonplace this century on both sides of the Atlantic: Chevron and Total are comparably eager in their greening. Moreover, according to Saeverud and Skjaerseth (2007) there has been consistency between what the big three (BP, ExxonMobil and Shell) say about climate change and what they do.

Since 2003 there has been a change at the top in both BP and ExxonMobil: how do those two former protagonists on climate change stand on this matter under new leadership? What about the rest of the oil industry, most of it state-controlled? How much are oil companies constrained in what they can do? These matters will now be examined in Chapter 4.

4

What is the oil industry supposed to do?

4.1 THE CHOICE OF STRATEGY

The historical scene has now been set, with the previous chapters of this book considering crucial earlier events, both 55 million years ago and also during the last few decades. Now is the point at which we move away from the specifics of geological history and the past fortunes of individual companies, and begin to consider the broad choices facing the oil industry during this century, as the end of the Age of Oil approaches.

The twentieth-century global economy was based significantly on energy produced by burning hydrocarbons with unfettered release of the resulting carbon dioxide into Earth's atmosphere. Note that word hydrocarbons. This is far from being just to do with oil and gas: the use of coal is central to the matters under discussion here. The twenty-first century has begun in the same fashion as the twentieth ended, with similar reliance on hydrocarbons (Figure 4.1) and with similar variation in regional patterns of consumption of hydrocarbons (Figure 4.2). How can the oil industry best move away from that established position, should it wish to do so?

For individual companies the choice of route and timing will clearly be much more specific than for the industry as a whole. At one extreme a particular company could decide to access and harvest the remaining oil and gas and then wind up the business. With that strategy in place, timing of close of play would be determined by the moment at which some combination of supply and

67

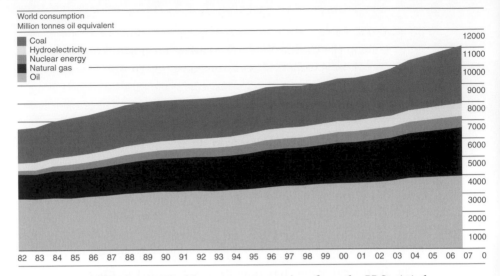

Figure 4.1 World energy consumption, from the *BP Statistical Review of World Energy* (BP, 2008). Coal is the fastest-growing fuel.

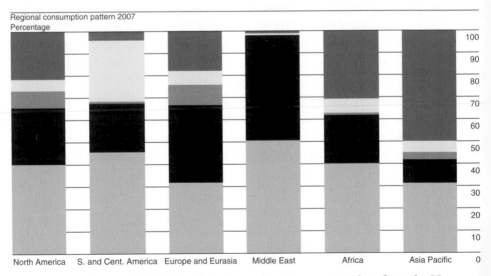

Figure 4.2 Regional patterns of energy consumption, from the *BP Statistical Review of World Energy* (BP, 2008). Coal dominates in Asia Pacific: it meets 70% of China's energy needs. Key as for Figure 4.1 (above).

demand regularly reduced profits below a commercially acceptable level. At the other end of the spectrum, a company could aim to be a leader in making the transition to a low-carbon economy. The rate at which that change would be made would be determined by the extent to which governments and stakeholders expressed in practical terms a conviction that such a course of action should be made commercially viable.

This is not to say that the industry should be passive in the face of that choice. Conviction that a change is needed will be essential for there to be any hope of that taking place. A stark way of putting this is to ask the oil industry: do you wish to be defensive or constructive? I put such a challenge to my fellow members of the American Association of Petroleum Geologists (AAPG) through a letter in the house magazine *AAPG Explorer*, in February 2007:

> Lee Gerhard and I debated climate change recently in the *Bulletin* (March 2006). One point of agreement between us was the primacy of observational science, not least geological evidence. In his letter to November's *Explorer*, Lee, writing with Bill Pollard and Ray Thomasson, again asserts the value of geological observations. But they then go on to base their arguments against human-induced climate change on climatological rather than geological data ...
>
> *[There followed a brief rehearsal of the evidence from the 55 Ma event covered in* Chapter 2 *of this book.]*
>
> ... I believe that the AAPG has a particular responsibility to help the oil industry come to terms with the risks involved in our main activity of transferring fossil carbon from the ground to the atmosphere. Rather than fighting last-ditch battles using data largely derived from other sciences, the leadership of AAPG should accept the message from the rocks. Of course the world continues to need oil and gas. It also needs our leadership in capturing fossil carbon after use and storing it safely back where it came from, in the rocks we know so well. Our industry is at a critical point in considering climate change. Are we going to be defensive and moribund? Or constructive and vital?

AAPG President (2008–2009) Professor Scott Tinker wants us to be constructive and vital. He concludes his *Discussion Needs Climate Change* (*AAPG Explorer*, January 2009) with a clear call: 'Let's help lead.'

The public strictures from me on the behaviour of my colleagues invite the rejoinder that in reality the position is greatly complicated by the constraints under which the oil industry operates. The notion that the oil industry needs help from governments runs counter to traditional feelings inside non-state oil companies: the idea that this government help comes in the form of regulation might soften the view of the general populace towards both government and oil companies while running counter to deeply ingrained historical attitudes within the industry itself. Yet, in relation to the long-term strategic issue of transition to a low-carbon economy, it is indeed the case that government intervention is essential. It is even more obvious that the industry must retain the support of its financial backers and its customers.

Then there are the constraints imposed by Earth itself and its geological history. These constraints are implacable – not for nothing is an adamant individual compared in hardness with a diamond or otherwise described as granite-like – but they are constraints that can be understood and hence managed intelligently. That involves a certain type of behaviour.

To understand properly the strategic choices now facing the oil industry it is necessary to consider the rather unusual mixture of behaviour required of the active participants in that industry. On the one hand they have to do business with people, just like everybody else. Yet unlike most commercial activity, the oil business is based fundamentally not only on people but also on rocks. Unlike people, the original condition of those rocks cannot be changed: a point made to financial backers with some passion by geologists threatened with dismissal following a setback in exploration. You can indeed remove the exploration managers, or send them on behavioural management courses in an attempt to make them a little less enthusiastic about spending your money to uncover Earth's secrets. But you can't thereby alter the history of a gas-bearing sandstone lying beneath the bed of the sea off the shores of Trinidad, or that of an oil-bearing limestone buried deep below the deserts of Abu Dhabi. The behaviour required to deal successfully with oil exploration and production, particularly in the search for new ventures in frontier areas, involves a strange and coordinated mixture of qualities. The nature of that integration of tribal qualities is poorly understood outside the industry – and sometimes within the oil companies themselves. The industry has a fascinating mixture of problems to solve and opportunities to seize: it is already engaged in what is frequently an intellectually demanding task. It is a theme of this book that a positive response from the oil industry to its carbon challenge has great potential to help us manage the transition to a low-carbon economy. It is of real practical significance to us all.

4.2   THE CONSTRAINTS IMPOSED BY PEOPLE

There are constraints on both the state oil companies, who now control by far the greater part of the world's oil and gas reserves, and also on the international non-state operators such as ExxonMobil, BP, Shell, Chevron and Total, who at present control just a few per cent of those reserves – an apparently continuing reversal of the mid-twentieth-century position when Exxon, Shell, BP, Gulf, Texaco, Mobil and Chevron (the 'Seven Sisters') had ready access to much of the world's oil. One group is beholden to the nation–states that control them, the other to their main stockholders. On the face of it the constraints are unlikely to slacken soon for either group, with maybe the tightest grip exercised by the stockholders. Dr Chris Gibson-Smith, Chairman of the London Stock Exchange and a former director of BP, set out his view on UK plans to move to a low-carbon economy in an article (2007) in the *Sunday Telegraph* newspaper: 'We're hurting Britain, not saving the planet.' How could we get to the point where governments combine with the financial powers in the City of London and on Wall Street to urge oil companies across the world to diversify into low-carbon energy supply, rather than simply harvest the remaining supplies of oil and gas?

Could energy companies and governments alike move swiftly away from dependence on fossil fuels? The short answer is no. A major reason for this reluctance to change is the affection with which the immediate benefits of oil, gas and coal-generated electricity are regarded by the population at large. Customers of oil companies, also voters in elections, find fossil fuels mighty useful, especially for personal transport. Even if they did not, they may be unconvinced that they are endangering the planet. Early research reported by Dr Clair Gough and Dr Simon Shackley of the Tyndall Centre for Climate Change Research (part of a UK consortium), Manchester University, at the *Coping with Climate Change* meeting in 2003 (Chapter 3), showed extensive denial of anthropogenic climate change in Britain, not least among groups with access to good information. Present-day opinion polls suggest that there is still a notable lack of conviction. On 22 June 2008 *The Observer* newspaper reported data from an Ipsos MORI poll which showed that:

> The majority of the British public is still not convinced that climate change is caused by humans – and many others believe scientists are exaggerating the problem …

Oil companies, in common with other types of organisation, are increasingly prepared to publish records of their emissions of potential pollutants during their normal operations and to announce and achieve targets to reduce these emissions. This concern with clean operations is quite a different matter from any one oil company refusing to continue to take fossil fuels from below the surface of the planet, and thereby ceasing to be a prime agent in recycling carbon into Earth's atmosphere at a far greater rate than that which would otherwise obtain.

However environmentally enlightened an oil company might be, there would still be a major non-geological, indeed non-scientific, constraint on any retreat from fossil fuels: the difficulty of maintaining essential profitability during that retreat. For the non-state oil companies, a crucial potential veto would be exercised by holders and prospective purchasers of that company's shares. Could such an early-twenty-first-century oil company pursue even a medium-term strategy explicitly aimed at reducing profit based on production of fossil fuels, without cutting the main sources of income and thereby causing the extinction of the organisation? Successful implementation of such a strategy would require a remarkable combination of scientific understanding and technology, plus the inculcation of an educated awareness of certain unfamiliar and long-term issues within the investment community. Such integrated activity may be some years away, barring concerted efforts by the oil companies themselves.

The emphasis would have to be on concerted action. Simple withdrawal from production of oil and gas by major oil companies would undoubtedly lead to massive disruption in economic development in many countries. Withdrawal would also create opportunities for companies less concerned with environmental matters to step in and seize the opportunities thereby created to take up acreage, facilities and customers. Such organisations would undoubtedly include national oil companies, which are likely to remain relatively free of the sort of pressure from shareholders and environmental groups habitually experienced by the likes of ExxonMobil, Shell and BP.

The state oil companies, the ten largest of which control some 70% of Earth's reserves of oil and gas, do of course have their own stakeholders: national governments. These stakeholders would be hard to convince of the need for movement away from the core of their business: not least the governments of such countries as Iran, Iraq, Kuwait, Saudi Arabia and the United Arab Emirates. Angola, Brazil, Nigeria and Venezuela provide further examples of countries where governments would be most reluctant to eschew the wealth generated by oil (Figure 4.3). These constraints appear to be unlikely to slacken as the state oil companies develop a stronger and stronger corner in remaining reserves.

So although it might be argued that a first step in controlling the dumping of fossil carbon into the atmosphere should come from the likes of ExxonMobil, Shell and BP, that is only the very beginning of the process. Those organisations are based in countries that have not got state oil companies. That has always been true of the USA and has been the case in the UK since the 1980s: the privatising Conservative government led by Margaret Thatcher was first elected in 1979. Large discoveries in Alaska and the North Sea gave the USA and UK oil industry some late-twentieth-century release from the extensive state control elsewhere that followed earlier nationalisation of the industry's assets in Nigeria, Libya and the Middle East. But production from such giant fields as Prudhoe Bay, Forties, Ninian and Brent is now past its peak. As the new century starts, nationalisation – explicit

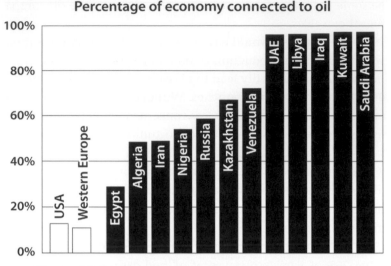

Figure 4.3 Dependence of Saudi Arabia and various other countries on oil, from Gautier (2008, her Figure 7.5). Although the USA here appears less vulnerable, it might not seem that way in Washington, DC (see comments in 2005 of President George W. Bush, reported in Chapter 7).

in Venezuela and slightly cryptic in Russia – has further constrained the freedom of action of the independent international oil companies.

4.3   THE CONSTRAINTS IMPOSED BY ROCKS

Beyond organisations lies something man cannot change: only part of Earth was covered by the ancient Tethys Oceans (Figure 4.4). Earth's abundant and easily accessible reserves of oil and gas are disproportionately located in the sedimentary rocks formed around the margins of those ancient seas (Table 4.1, Table 4.2). That is because oil and gas accumulate preferentially on the margins of the continents rather than in continental interiors – and do not accumulate at all in commercial quantities over the greater part of Earth's surface that is covered by the oceans and is remote from the continents.

Oilfields do not last for ever. Some have survived for hundreds of millions of years in the most stable parts of the

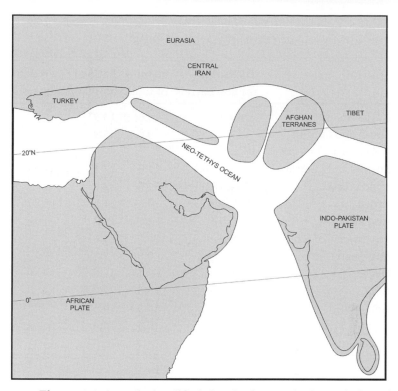

Figure 4.4 A much-simplified sketch of a stage in the evolution of the Tethys Oceans, after Sharland *et al.* (2001). Here 'Neo-Tethys' Ocean is shown during its closure from 63 to 34 million years ago. The fossil carbon in the oil and gas formed in this region would have been naturally recycled into Earth's systems over tens of millions of years had we not intervened and released it over a couple of hundred years.

continents, but the general life expectancy of a deposit of oil or gas would be 100 to 200 million years – assuming that the reservoir has not previously been drained by us. As the tectonic plates move around the surface of the planet at the rate of a few centimetres a year, so oilfields will be destroyed as plates converge and crunch the intervening layers of rock, rupturing the seals on the oil and gas reservoirs.

On an inhumanly long timescale fossil fuels are indeed renewable. Two hundred million years from now, a form of life requiring abundant oil for some purpose should find that plenty has formed since our own times. Present-day geologists can with

Table 4.1 *Proved reserves of oil by region at end 2007. Data from* BP Statistical Review of World Energy *(BP, 2008).*

	Thousand million barrels	% share of total	Ratio of reserves to annual production
North America	69.3	5.6	13.9
South and Central America	111.2	9.0	45.9
Europe and Eurasia	143.7	11.6	22.1
Middle East	755.3	61.0	82.2
Africa	117.5	9.5	31.2
Asia Pacific	40.8	3.3	14.2
Total world of which:	1237.9	100	41.6
European Union	6.8	0.5	7.8
OECD	88.3	7.1	12.6
OPEC	934.7	75.5	72.7
Non-OPEC	175	14.1	14.3
Former Soviet Union	128.1	10.4	27.4
Canadian oil sands	152.2		

Table 4.2 *Proved reserves of natural gas by region at end 2007. Data from* BP Statistical Review of World Energy *(BP, 2008).*

	Trillion cubic metres	% share of total	Ratio of reserves to annual production
North America	7.98	4.5	10.3
South and Central America	7.73	4.4	51.2
Europe and Eurasia	59.41	33.5	55.2
Middle East	73.21	41.3	Over 100
Africa	14.58	8.2	76.6
Asia Pacific	14.46	8.2	36.9
Total world of which:	177.36	100	60.3
European Union	2.84	1.6	14.8
OECD	15.77	8.9	14.4
Former Soviet Union	55.53	30.2	67.7

impunity speculate about where that might happen. I have always quite liked the deep, central Black Sea as a contemporary frontier exploration venture, and was involved in acquiring a licence for BP to operate there in the 1990s. This area is yet to be tested in the area of highest geological risk in very deepest waters. In November 2008 ExxonMobil and the Turkish state oil company TPAO announced a new agreement to explore the deepwater Black Sea. Even if present-day drilling by ExxonMobil/TPAO were to be unrewarding, I should still happily bet on success after another 100 million years has passed and the carbon-rich black muds on the floor of the Black Sea have been converted on further burial into a rich source rock for oil.

But exploration for oil and gas is about recognising what happened in the past, not predicting the future. In many places rich in hydrocarbons there is readily available evidence at the surface of what lies below. Even following the desecration of the great buildings at Babylon by modern bricks, egotistically inscribed by Saddam Hussein with his name coupled with that of Nebuchadnezzar, it is possible to see clearly the natural bitumen cementing the earlier building (Figure 4.5). The burning fires of Babylon have their gas-fuelled cousins in the eternal flames on the flanks of Kirmaky valley north of Baku, Azerbaijan (Figure 4.6). A queasy walk can be taken on the yielding surface of Pitch Lake, Trinidad (Figure 4.7). These more obvious hydrocarbon provinces were discovered and exploited early in the history of the oil industry. They still hold reserves and exploration is by no means complete.

Yet Earth is implacably finite. Even though new ventures are made possible by technology that leads exploration and production into deeper and deeper parts of the ocean, and into polar regions, such ventures cost an order of magnitude more than extracting oil from beneath the deserts of Arabia. So does mining oil from tar sands and oil shales, even without costing the environmental damage caused by the use of large volumes of gas and water to liberate the sticky oil from the sands (Cope, 2008, 2009). While there continues to be a vigorous debate about the timing of the peak in global oil production, it is beyond dispute that the distribution of the remaining reserves, determined millions of

Figure 4.5 Bitumen used to cement Babylon bricks (right) and ancient inscription on those bricks. Saddam Hussein built over many of the earlier structures using contemporary cement and slogans. Photograph by the author (coin is 25 mm diameter).

Figure 4.6 Eternal flames, Yanadarg, Azerbaijan. Photograph by Miss Claire Budd. © Claire Budd.

Figure 4.7 Natural deposit of hydrocarbons at Earth's surface at Pitch Lake, Trinidad. Balancing on the surface is a group of participants in BP's *Challenge* early development programme. © Claire Budd.

years ago, is a first-order constraint on the strategic choices now available to the oil industry at large.

There is not currently much debate about the peak in global coal production, because of the abundance of remaining reserves. Nor is there currently much conflict over access to areas of coal production, as there has frequently been over access to areas of oil production. Reserves of coal are spread more evenly between nations than are reserves of oil. This relative ease of access to coal compared with oil presents the likelihood of massive additional releases of carbon dioxide into the atmosphere from burning that coal. This threat should constrain severely the use of coal to generate electricity from large power stations, unless those releases can be captured and stored (Rees and Butler, 2008). Fortunately the use of coal in power stations does offer an opportunity to capture and store the emissions of carbon dioxide safely underground: low-carbon electricity from an abundant fossil fuel. Can this process be a key part of a safe and economic transition away from the use of fossil fuels?

Is this a matter in which oil companies should get actively involved? Carbon capture and storage (CCS) is a different business from oil exploration and production: in the simplest terms it is about waste management. It is argued in this book that the oil companies could themselves play a major role in CCS, using the many skills in the existing workforce that can readily be applied to the task. An alternative view (personal communication, Mr Nick Butler, Judge Business School, Cambridge University, 2009) is that this expertise might be best seconded to new business organisations under government direction. This matter is discussed further in Chapter 7. A clear point of agreement between Butler and me is that the skills in understanding rocks and fluids that are essential for successful CCS operations already exist within oil companies. The constraints on the subsurface storage of carbon dioxide are familiar topics to reservoir engineers and reservoir geologists (Chapter 6).

4.4   RESPONDING TO THE CONSTRAINTS

What is the attitude of the state oil companies, the organisations that control most of the reserves of oil and gas on Earth? Dr Valerie Marcel of Chatham House (a London-based non-profit non-governmental organisation that promotes understanding of international affairs) examined the strategic priorities of the National Oil Companies (NOCs) in a series of interviews, a summary of which has been published by KPMG International (2008). She interviewed executives from Saudi Aramco, Gazprom (Russia), National Iranian Oil Company (NIOC), Iraq's Southern Oil Company (SOC) and Iraq National Oil Company (INOC), Kuwait Petroleum Corporation (KPC), China National Petroleum Corporation (CNPC), PetroChina and China National Offshore Oil Corporation (CNOOC) and Petrobras (Brazil). Based on these interviews and other research, she found that the main concerns were with such issues as lack of skilled personnel, rising costs, political instability and declining reserves, with climate change and environmental regulation languishing some way down the list of perceived risks to business.

The KPMG report states (2008, p. 17):

> Climate change does not yet figure very highly on the strategic horizon for a number of NOCs. 'No impact. Climate change could raise the operation cost, but it is under control', said a Chinese executive. At Petrobras, Milton Costa Filho saw the strategic risk from climate change as affecting their business in 20–30 years' time, not in the coming decade.

Yet the KPMG report concludes that oil companies that have invested in the development of technologies and markets for low-carbon energy could have a strategic advantage in the coming decade: 'National oil companies that have not addressed these new challenges internally may need to turn to partners for help.'

One state oil company that may feel no need of such help is Norway's Statoil, now StatoilHydro. A tax on release of carbon to the atmosphere imposed by the Norwegian government made it commercially advantageous for Statoil, with its partners ExxonMobil, Hydro (then a separate company) and Total, to store carbon dioxide co-produced with natural gas from the Sleipner field in the North Sea rather than vent it to the atmosphere. The natural gas at Sleipner contains around 9% carbon dioxide, well above the limit for commercial gas quality: the carbon dioxide has to be captured in any event. It is the storage that is special. This pioneering operation injects approximately 1 million tonnes of produced carbon dioxide a year into the Utsira Formation 'saline aquifer' – a rock with salty water in its pores – a kilometre below the bed of the sea (Fredriksen and Torp, 2007; see also Chapter 6).

Chevron has plans for injecting carbon dioxide co-produced with natural gas in their Gorgon project on Barrow Island, off the northwest coast of Australia. Over the life of that project approximately 120 million tonnes of carbon dioxide is expected to be safely stored over 2 kilometres underground. Meanwhile, a million tonnes a year of carbon dioxide is already being captured and stored at In Salah, Algeria, by BP and partners Sonatrach and Statoil. As at Sleipner, this is to deal with carbon dioxide co-produced with natural gas. At In Salah the carbon dioxide is re-injected into the same geological formation from which it originally came, but at a deeper level, below the gas–water contact. Unlike the position at Sleipner, there is no obvious commercial benefit attached to this project, because there is no regulatory framework providing for long-term underground storage of carbon dioxide. So for BP and partners this is a loss-making enterprise, costing about $6 per tonne of carbon dioxide not released to the atmosphere (Nicholls, 2007a). BP and partners might reasonably regard this as an investment for a future in which carbon allowances make such ventures commercially feasible as well as environmentally desirable. At Sleipner the value of avoiding the release of a tonne of carbon dioxide was around $55 in 2007 (Fredriksen and Torp, 2007): a figure to yearn for at In Salah. (These matters are considered in more detail in Chapters 6 and 7.)

Capture and storage of carbon dioxide featured in the 2003 debate in London between BP and ExxonMobil (Chapter 3). The debate itself was a clear public demonstration that the challenges of climate change were at least being considered by the leaders of the independent (non-state) oil companies on both sides of the Atlantic. That exchange between BP and ExxonMobil took place with John Browne and Lee Raymond well established as leaders of those two companies. Subsequent changes at the top of both companies have been followed by speeches by the new leaders that give a measure of convergence – or otherwise – in publicly stated policies on climate change. These speeches may be compared with the policies set out so clearly in Burlington House, on that March evening in 2003 when the most recent Iraq war was illustrating a whole range of constraints on oil-industry strategy (Chapter 3). The following extracts from the 2007 statements of policy suggest that there remained a difference in the degree of public commitment by the two companies. In a speech in Houston in February 2007, Mr Rex Tillerson (Figure 4.8) of ExxonMobil had this to say on the topic 'Addressing emissions' in a review of the state of the energy industry (Tillerson, 2007):

> No discussion about the realities facing our industry today would be complete without reference to the issue of greenhouse gas emissions and climate change. This is an issue that crosses all boundaries, impacts industry and governments, but most importantly will directly impact consumers in every part of the world.

Figure 4.8 Mr Rex Tillerson, Chairman of ExxonMobil, who inherited an officially sceptical corporate line on anthropogenic climate change from his predecessor, Mr Lee Raymond (see also the comments of Frank Sprow at the BP–ExxonMobil debate of 2003, discussed in Chapter 3). Photograph from ExxonMobil.

The majority of the growth in energy demand will come from developing nations as their growing populations pursue higher standards of living. With this improvement in living standards will come most of the growth in future greenhouse gas emissions.

By the year 2030 it is expected that global emissions of carbon dioxide will approach 40 billion tons per year, up from close to 28 billion tons per year today.

So, we know our climate is changing, the average temperature of the earth is rising, and greenhouse gas emissions are increasing. We also know that climate remains an extraordinarily complex area of scientific study. While our understanding of the science continues to evolve and improve, there is still much that we do not know and cannot fully recognize in efforts to model and predict future climate system behavior.

Having said that, the risks to society and ecosystems from climate change could prove to be significant. So, despite the uncertainties, it is prudent to develop and implement sensible strategies that address these risks while not reducing our ability to progress other global priorities such as economic development, poverty eradication and public health.

Our industry has a responsibility to contribute to policy discussions on these important issues – and to take concrete actions ourselves to reduce emissions.

As an industry, we are already improving efficiency in our operations – greatly enhancing our energy efficiency while supplying more products than ever before. Steps taken at ExxonMobil, for example, since 1999 to improve energy efficiency at our facilities, for example, resulted in CO_2 emissions savings of 11 million metric tons in 2005. That's equivalent to taking two million cars off the road.

But we must do more. We must continue to foster and support scientific research into technology breakthroughs to deliver new sources of energy with even lower emissions. One example is Stanford University's Global Climate and Energy Project, which ExxonMobil and other partners are supporting with a collective contribution of $225 million.

The approaches policymakers adopt to address climate risks are also important. A global approach is needed that promotes energy efficiency, ensures wider deployment of existing emissions-reducing technologies and supports research into new technologies.

It is also critical to maintain support for fundamental climate research, recognizing that there remains much that we still do not understand. Specific policy tools should be assessed for their likely effectiveness, scale, and costs, as well as their implications for economic growth and quality of life. In that regard, rigorous and informed debate – debate that takes into account the essential role played by energy in advancing social and economic progress – will best support thoughtful policymaking.

In our view, the most effective approaches will maximize the use of markets. This will help promote global participation and facilitate the rapid spread of successful initiatives.

Consistent with a market-based approach, effective policies will ensure a uniform and predictable cost of reducing carbon emissions, maximize transparency, minimize complexity, and adjust to new developments in climate science and the economic impacts of policies.

Just as technology has continually been the driver of progress in our industry, I am confident that future technology advances will both expand our understanding of the climate system and enable an effective response.

We must encourage all participating in this debate to frame the discussion in terms of the realities we face – the realities of growing demand and the need for affordable, reliable energy to enable the world's consumers to achieve genuine improvements in their quality of life.

The policy measures adopted today will have far-reaching implications in the years ahead. We must consider the potential impacts on future economic growth and quality of life for not just the current generation, but those of our children and grandchildren.

Compare this 2007 view of the new leader of ExxonMobil with an opinion from the top of BP, given in a speech made in Berlin by Dr Tony Hayward (Figure 4.9) soon after his appointment as Chief Executive. The topic was 'Delivering technologies via carbon markets' (Hayward, 2007):

In conclusion, I would like to say:

Now is the time for us all to roll up our sleeves and take the necessary action on climate change

We already have the technology to make a difference and companies like BP are already making some progress

Well functioning markets are preferable to subsidies

If a market price for carbon is created, it will generate a cascade of incentives throughout the economy which will encourage massive emission reductions

Transitional incentives should be put in place, which will accelerate the cost reduction of new technology

And finally, global market is ultimately desirable, but we should push on with developing regional markets in the interim

I am an optimist; you have to be in business. Although the challenges of climate change are grave, they can be solved. Human ingenuity knows no bounds – and human kind will solve its biggest challenge to date.

Publicly BP appeared to continue to be more committed to taking action on climate change than was ExxonMobil, as was the case back in 2003. But from remarks made by Tillerson at Stanford University on 17 February 2009 it appears that any remaining gap is dwindling. The main differences now have to do with styles of regulatory framework within which the common goal of a low-carbon economy may be achieved.

Figure 4.9 Dr Tony Hayward, who succeeded John Browne as Chief
Executive Officer of BP in 2007. Hayward joined the company in 1982
as an exploration geologist, having developed his skills as an
observational scientist during fieldwork for his Ph.D. at Edinburgh
University. He is therefore particularly well placed amongst
contemporary leaders of the international oil industry to understand
the fundamentals of climate change. Photograph by Ted Nield.

Putting policy into effective practice is another matter from
issuing edicts to government on regulation. Which of the two
companies is the more likely to develop profitable forms of low-
carbon transport in collaboration with the automobile industry?
Which is the more likely to make substantial profits from ventures into solar power? From wind power? From storing carbon
dioxide safely underground? More crucially, will either BP or
ExxonMobil come to see these as areas of serious commercial
competition, comparable to hydrocarbon exploration and production? What of the other large oil companies? Shell (Spence,
2008) is certainly keen on carbon capture and storage. As so often
in the history of the oil industry, a large part of the answer lies
with the behaviour of national governments.

Ventures such as storage of carbon dioxide that are, or might
be, undertaken by oil companies are currently critically dependent
on the encouragement of national governments (in Australia, UK,
USA) and of state oil companies (in Abu Dhabi, Algeria, Norway).

Some of these projects aim to re-inject carbon dioxide co-produced with useful hydrocarbons in the course of the company's own operations. Strategically – but not environmentally – this is qualitatively different from operations that aim to store the carbon dioxide that is captured following human use of those hydrocarbons: the oil companies thereby taking a measure of responsibility for their customers' use of their products (see the quotation from John Browne's 1997 speech in Chapter 3).

By early 2008 considerations of the relatively high price of oil, and of energy security for those living outside OPEC (Organization of Petroleum Exporting Countries) and Russia, were driving both government policy and the international oil industry in contradictory directions. Environmentally undesirable extraction of oil from tar sands, and expensive frontier exploration in Arctic waters, were commercially approved by the non-state oil companies with a readiness barely conceivable only a few years ago. These companies were now more confident of a decent return on the large capital sums employed: a confidence that slackened as the price of oil dropped sharply during the global financial crisis later in 2008. These actions – be they commercial or not in the outcome – do nothing to reassure those who think that any greening of the industry until now has been largely cosmetic.

In contrast to those efforts to prolong the traditional activities of the oil industry, are government-led initiatives to promote low-carbon alternatives. These initiatives look much better financially with oil at $50–100 a barrel rather than $10–20, but cannot be contingent on the often volatile price of oil, or indeed on changes in the wider financial background such as those felt across the world in the autumn of 2008.

A clear message from the rocks, echoed by some governments, is now telling the petroleum engineers and geologists that the dumping of carbon dioxide into Earth's atmosphere by use of fossil fuels is a problem. A clear message from the market, not least the customer, is telling these same oil folk that their main product is still highly valued. In the face of those apparently conflicting pressures, all the oil companies have significant problems in common. In the light of that, historical differences between companies begin to fade into the background. Together, what are they supposed to do?

4.5   WHAT IS THE VALUE OF VIRTUE?

If growth in environmental virtue in the oil industry is not encouraged by governments in Beijing, Delhi and Washington, and financial support for that move is not available in London and New York – so what? Does it matter whether the oil companies are involved significantly in developing a low-carbon global economy?

How can the leading non-state oil companies such as BP, ExxonMobil and Shell become environmentally virtuous? Little more than 10% of the emissions of carbon resulting from their activities arises directly from their own operations; the remainder, approaching 90% of the total, comes from the use of their products by their customers. We have seen in the previous section of this chapter that ExxonMobil has a commercial incentive to store carbon dioxide at Sleipner, because of a tax on released-carbon imposed on their StatoilHydro-led partnership by the government of Norway. In contrast, BP and partners are taking a loss at In Salah in Algeria with a comparable operation, while preparing for a possible future in which Norwegian-style regulatory frameworks are a standard part of the operations of oil companies.

Even if Sleipner-style incentives to environmental virtue in operations were to become widespread, that would only address a small part of the problem. Some 90% of the carbon emissions resulting from the activities of oil companies come from their customers, not the companies themselves. The companies' products are considered essential to economic and social well-being by their shareholders, who are also their customers. Those shareholders are unlikely to include many of the entrenched opponents of the oil companies; opponents who would argue that the sooner an oil company went out of business the better.

At least one flaw in this argument is the threat it carries for potentially crucial allies of the environmentalists within the industry. In the absence of a good regulatory framework, oil companies that strive for low-carbon virtue become more vulnerable financially in at least the short term than those who simply pump oil and gas. Big Oil might accept the reality of anthropogenic climate change, and yet predict that fossil fuels will form an essential part of energy provision through to the middle of this century and beyond (see the 2003 BP–ExxonMobil debate, Chapter 3).

To some such passivity in the face of that paradox is unacceptable. If convulsions in the energy market are required to prevent the catastrophe of unchecked climate change, then let it be. A new and better order can emerge. These lofty and high-level arguments are unlikely to prevail in either rich or poor countries. Both those in developed countries, who have enjoyed the great material comfort and personal convenience brought by the products of the oil and coal industries during the twentieth century, and those in developing nations who aspire to such benefits in the twenty-first century, will need to be convinced that the rapid elimination of those industries is really necessary. They will get no such conviction from this book.

In the next two chapters I shall set out the technical and other reasons why I believe we can continue to have our fossil-fuel cake and eat the more nutritious parts of it without causing ourselves serious damage. But we will have to change our behaviour. This change will not necessarily involve much physical discomfort, indeed it may lead to a real improvement in our lives. But it will involve something at least as tricky, which is managing a high level of intellectual challenge to the existing order. We have to develop widespread conviction that we face a huge problem and that by great resolution we can tackle that problem.

For that we need help from the oil industry. We need a full and deep understanding by the oil industry of two main things. The first is the message from the rocks that the industry's activities have played a significant part in creating the problem in the first place. The second is the recognition that the expertise of the industry can play an equally significant part in helping to solve that problem: by storing safely underground carbon dioxide captured from coal-fired power stations. This expertise is required urgently (Rees and Butler, 2008). We will still need fossil fuels at least to mid-century as we make the transition to other forms of energy, but we cannot safely release to the atmosphere the carbon dioxide resulting from their combustion.

We all find ourselves in an unusual situation. If we respond with our normal tribal impulses, we shall fall well short of what is required. We need the oil industry to help us manage the transition from our current great dependence on its products to a low-carbon future. For that to happen, governments have to take an uncharacteristically long-term view and establish an appropriate framework (Chapter 7). What can the oil industry offer to encourage that government action? That question is addressed in Chapters 5 and 6.

5

The size of the problem and the scale of the answer

5.1 THE PRINCETON WEDGES

The environmental movement in the 1990s insisted that the oil industry must greatly curtail its activities if we were to avoid serious damage to the planet. Like any spirited child when confronted by the advice of granny that you can't have your cake and eat it, the industry's reaction has been a determined attempt to see if granny is right. In debates organised as a centrepiece of oil industry training programmes around the turn of the century, I would hear pleas from senior company folk to their environmentalist opponents: 'If carbon is the problem, just tell us, and we will solve it.' This belief in a technological solution would habitually infuriate many of the environmentalists, who interpreted it as yet another attempt by the oil barons to avoid taking real responsibility on behalf of their company.

If it were possible to continue to enjoy the use of fossil fuels for certain particularly convenient purposes, without causing environmental harm, that might be seen by many as a good deal. It is certainly well worth the oil industry's while to explore that possibility. Indeed stakeholders might believe it was essential that this possibility be examined simply to protect their interests.

Much has been made of the funding by the oil industry in the USA of research into climate change with the alleged purpose of proving that the climate-change sceptics are right. History will judge just how significant that may have been, but now any such work appears to be ephemeral. Frank Sprow spoke openly of the large investment in energy research made by ExxonMobil at Stanford University in the 2003 *Coping with Climate Change* debate (Chapter 3). His fellow debater Greg Coleman referred to BP involvement in comparable research, including work at Princeton University. This is the Carbon Mitigation Initiative, a ten-year programme launched in 2001 in which BP has invested $15 million and Ford $5 million. The initiative was, in October 2008, emphatically extended to 2015 (Eyton, 2008).

The Princeton research produced an early result, a paper in the influential journal *Science* by Professor Stephen Pacala and Professor Robert Socolow, 'Stabilization wedges: solving the climate problem for the next 50 years with current technologies' (2004). This article repays the closest attention. It makes the large claim, supporting the existing view of the Intergovernmental Panel on Climate Change, that we already possess the means to give ourselves a chance of solving the climate problem over the next 50 years. Beyond that things are at least as tough, but at least we shall have retained some measure of control. The advantage of looking at these matters over 50 rather than 100 years is that the shorter period of time has a real personal meaning, even for those of us who are grandparents already and are unlikely to last that long ourselves.

Here is the Abstract of 'Stabilization wedges':

Humanity already possesses the fundamental scientific, technical, and industrial know-how to solve the carbon and climate problem for the next half-century. A portfolio of technologies now exists to meet the world's energy needs over the next 50 years and limit atmospheric CO_2 to a trajectory that avoids a doubling of the preindustrial concentration. Every element in this portfolio has passed beyond the laboratory bench and demonstration project; many are already implemented somewhere at full industrial scale. Although no element is a credible candidate for doing the entire job (or even half the job) by itself, the portfolio as a whole is large enough that not every element has to be used.

Pacala and Socolow come to the issue from different angles, at least formally: Pacala from the Department of Ecology and

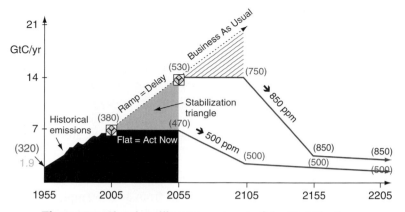

Figure 5.1 Sketch to illustrate concept of the 'Stabilization triangle', as conceived by Pacala and Socolow (2004, 2006), indicating the scale of the task involved in holding levels of carbon dioxide in the atmosphere to 500 ± 50 parts per million by the middle of this century. Vertical axis shows gigatonnes (thousand million tonnes) of carbon released each year (horizontal axis). If we continue 'business as usual', 20 gigatonnes of carbon will be emitted each year by the end of the century. If we act now, emissions can be held at the early twenty-first-century level of 7–8 gigatonnes a year through to mid-century. Then even more drastic action will be required to reduce atmospheric levels of carbon dioxide to what is generally considered to be a safe level. The 'Stabilization triangle' represents the difference between acting now and business as usual. It can be considered as a series of wedges – here termed the 'Princeton wedges' – that form a basis for effective action (see Table 5.1). The original number of wedges required for stabilisation was seven. Continuing increase in emissions means that eight wedges are now required (see Figure 5.5). Note that in Earth's atmosphere 1 part per million of carbon dioxide is equal to 2.1 gigatonnes of carbon. Slide shown by Professor Robert Socolow at the Geological Society, London on 12 October 2005.

Evolutionary Biology, Socolow from the Department of Mechanical and Aerospace Engineering. They introduce the concept of the 'stabilization triangle' (Figure 5.1), originally consisting of seven 'wedges' of technology. Each of these wedges can contribute significantly towards resolution of the problem of controlling use of fossil fuels and amending agricultural practice (Table 5.1). Table 5.1 carries a most important cautionary message. We can indeed use existing technology to make a huge contribution

Table 5.1 *Fifteen ways to make a Princeton wedge (after Pacala and Socolow, 2004, 2006).*

Choice of options: original requirement was for seven wedges – now eight are needed because emissions have continued to increase.	Effort by 2056 for one wedge. Each wedge saves 25 gigatonnes of carbon when phased in over the next 50 years, building to 1 gigatonne a year by 2056.
End-user efficiency and conservation	
1. Efficient use of vehicles	Increase fuel economy of 2 billion cars from 30 to 60 mpg
2. Reduced use of vehicles	Drive 2 billion cars only 5000 miles a year rather than 10 000 miles a year
3. Efficient buildings	Cut use of electricity by 25% in homes, offices and shops
Power generation	
4. Efficient coal base-load plants	Raise efficiency at 1600 large (1 gigawatt capacity) coal-fired plants from 40% to 60%
5. Gas base-load power substituted for coal base-load power	Replace 1400 large coal-fired plants with gas-fired plants
Carbon capture and storage (CCS)	
6. Capture carbon dioxide at base-load power plant	Install CCS at 800 large coal-fired power plants
7. Capture carbon dioxide at hydrogen plant	Install CCS at coal plants that produce hydrogen for 1.5 billion vehicles
8. Capture carbon dioxide at coal-to-synfuels plan	Install CCS to capture half of the carbon in the coal used to produce 30 million barrels a day of synfuels
Scale of geological storage required for a CCS wedge?	***Create 3500 Sleipners, thereby storing 3.5 gigatonnes of carbon dioxide (1 gigatonne of carbon) a year***
Alternative energy sources	
9. Nuclear power for coal power	Add twice today's nuclear output to displace coal

Table 5.1 (*cont.*)

Choice of options: original requirement was for seven wedges – now eight are needed because emissions have continued to increase.	Effort by 2056 for one wedge. Each wedge saves 25 gigatonnes of carbon when phased in over the next 50 years, building to 1 gigatonne a year by 2056.
10. Wind power for coal power	Increased wind power 40-fold to displace coal
11. Photovoltaic power for coal power	Increase solar power 700-fold to displace coal
12. Wind-generated hydrogen in fuel-cell car for petrol in hybrid car	Increase wind power 80-fold to make hydrogen for cars
13. Biomass fuel for fossil fuel	Drive 2 billion cars on ethanol, using 1/6 world cropland
Agriculture and forestry	
14. Reduced deforestation plus new planting	Stop all deforestation
15. Conservation tillage	Expand conservation tillage to 100% of cropland

towards solving the problem of anthropogenic climate change, but we can only do so by applying that existing technology on an heroic scale. So, in delivering Wedge Eight, one of the requirements is to create 3500 storage sites for carbon dioxide each with the capacity of the pilot site at Sleipner in the North Sea (Chapters 4 and 6).

Before we consider the particular wedges identified by Pacala and Socolow, it should be emphasised that their 'headline' choices are not exclusive. For example, 'recycling' the large amount of heat currently wasted during the generation of electricity is not explicit in their list, but could make a considerable contribution. Casten and Schewe (2009, p. 30) consider the numbers for the United States: 'total recycled energy could be 200 gigawatts per year, equaling 20 percent of the power being generated in the U.S'.

Of the 15 potential wedges chosen for illustration by Pacala and Socolow (Table 5.1), most involve the fossil-fuel industries in one form or another. Companies such as BHP Billiton that mine minerals and coal as well as producing oil and gas could put a tick in just about every box were they considering a contribution. In principle an oil company could decide to invest heavily in one or more of the wedges, but there are serious practical constraints (Chapter 4). Frank Sprow of ExxonMobil made an unscripted intervention in the 2003 *Coping with Climate Change* debate, when on hearing of BP's struggle to make profits from its investments in solar power he commented:

> We like BP spending money on things that do not make much money.

It is a theme of this book that the oil companies can make a significant contribution to the Princeton targets, particularly in relation to the capture and storage of carbon dioxide emitted by coal-fired power stations. It may be a harbinger that Shell announced in March 2009 that its investment in carbon capture and storage (CCS) would continue, while new investment in wind and solar power would cease. Given the will by both oil companies and government, CCS is a process that could be implemented in time to meet the requirements of the Princeton schedule.

Princeton Wedges One, Two, Three and Four fall under the title 'Energy efficiency and conservation'. The oil industry gets constructively involved here at the cost of reduced sales of its products, which is on the face of it a quite remarkable example of taking responsibility for the use of its products.

Development of efficient vehicles involves the downstream oil industry, that part involved with refining and marketing, in collaboration with the motor industry. That partnership is already long-standing: the joint funding of the Princeton study by BP and Ford is an obvious example of its continuation into this century. Both increased efficiency and reduced use of vehicles have obvious implications for marketing of motor fuels. A 2002 study for BP led by Mr Francisco Ascui and colleagues at the Judge Institute of Management (now the Judge Business School) at Cambridge University looked at how that company might achieve its professed ambition of moving 'Beyond petroleum' in this area. Will the oil companies just lick their wounds or will they seek to get involved more than they are now in providing personal mobility rather than just fuel? Meanwhile BP offers its motoring customers the opportunity to buy carbon offsets to provide some degree of environmental absolution from their estimated annual mileage.

BP also provides a handy calculator on its website so households can estimate their annual output of carbon. For those of us who are habitual users of aeroplanes the figures are chastening. If many of us respond by spending more time in our own country, and travel through that by train, the reduction in carbon footprint will be significant. So presumably will be the reduction in the profits of BP Air as demand for aviation fuel declines. Further experiments with the carbon calculator will show why creating energy-efficient buildings is now widely recognised as one of the most satisfactory ways of reducing emissions of carbon dioxide. Providing energy to buildings accounts for 40% of the total demand for energy in the UK. In many cases existing buildings are highly energy-inefficient (Fitzgerald, 2005; Palmer et al., 2006). The scope for cutting demand for fossil fuels is considerable.

Princeton Wedge Five stands on its own as a shift in fuel use from coal base-load power to gas base-load power. Pacala and Socolow comment that 'competing demands for natural gas' are involved here. Those taking a longer view have for some time been concerned at the use of gas for generating electricity, even in modern efficient power stations. They regard gas as a relatively rare commodity that has a greater real value in the long run for other purposes: an example would be use as a chemical feedstock for the manufacture of a host of useful plastic objects. The oil companies have profited from the switch from coal to gas for generation of electricity that has already taken place. There has been another kind of profit for governments in countries where this has happened, because they have been able to claim low-carbon virtue as they strive to achieve their targets for reduction in emissions of carbon dioxide. Burning gas produces significantly less carbon dioxide per unit of energy produced than does conventional burning of coal, so in at least the short term the switch to gas is environmentally beneficial. Conversion

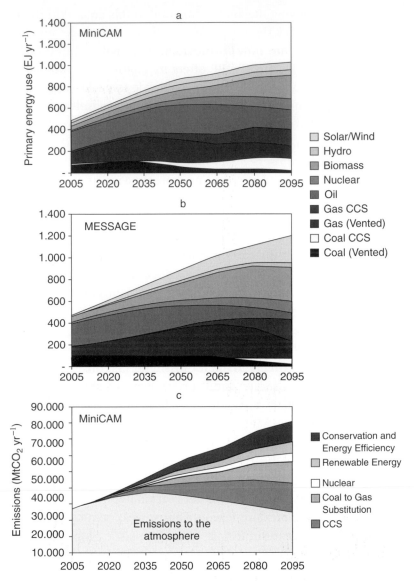

Figure 5.2 The contribution various activities might make to reducing emissions of carbon dioxide to the atmosphere, as seen by the Intergovernmental Panel on Climate Change (IPCC) (2005, their Figure TS-12). Two IPCC forecasts are shown, labelled 'MiniCAM' and 'MESSAGE'. The effect of substituting gas for coal is featured, and the role that can be played by carbon capture and storage (CCS) is emphasised (see Chapter 6). 'Mt' is a million tonnes. 'EJ' signifies an exajoule, 1 joule to the power of 18. *Wikipedia* notes that 1 joule is 1/100 of the energy a person can get from drinking a 'drop of beer'.

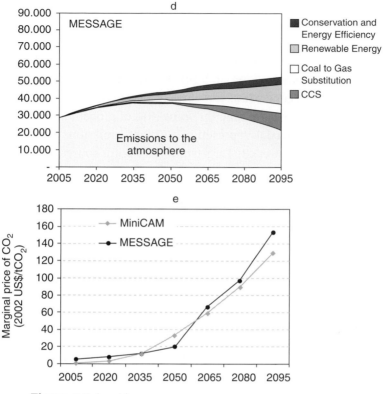

Figure 5.2 (cont.)

factors commonly used for the consumption of fossil fuels give a measure of this benefit: consuming oil yields 0.84 tonnes of carbon per tonne of oil, natural gas gives 0.64 tonnes and coal 1.08 tonnes per tonne of oil equivalent.

In some respects Wedge Five is the least satisfactory of the Princeton options: it is unlikely to make much of a dent in the longer run in the reliance of China and India on coal for generation of electricity. Environmentalists might reasonably claim that the switch from coal to gas for power generation offers the oil companies too easy a path to low-carbon virtue; if that is all the oil companies think they need to do. But the fact remains that the substitution of gas for coal can make a hefty contribution to reducing emissions (IPCC, 2005) (Figure 5.2).

Wedges Six, Seven and Eight are underpinned by another activity about which many environmentalists have long held a jaundiced view: the capture and storage of carbon dioxide. Here I suggest that the environmentalists are on thinner ice than in opposition to Wedge Five. Carbon capture and storage (CCS) offers a solution that is capable of being implemented with expertise already to hand, in a process where it is going to have a great effect: generating low-carbon electricity from coal. The power sector is responsible for some 40% of global emissions of carbon dioxide; coal represents c.70% of emissions from the power sector (Figure 5.3). Large stationary point sources emitting a lot of carbon dioxide (5–7 million tonnes a year for a 1-gigawatt plant) offer a good opportunity for significant CCS. Carbon capture and storage of this type is considered in detail in Chapter 6.

There is an intriguing possibility that a significant contribution to CCS may be made by a far less technologically dramatic approach, with which few environmentalists would wish to quarrel: the production of biochar – 'Black is the new green' (Marris, 2006). In this process, plant waste is burnt in low-oxygen conditions, producing both biofuel and charcoal. The carbon-rich char is returned to enrich the soil, thereby capturing carbon dioxide in quantities that it is claimed could make an immediate and significant contribution to a Princeton wedge (Glaser, Parr, Braun and Kopolo, 2009).

Wedge Nine concerns the substitution of nuclear power for coal power. James Lovelock has recently made the case for expansion in the production of electricity by nuclear fission, setting out with habitual vigour his change of mind on the issue in *The Revenge of Gaia* (2006). Yet in the eye of the public, the storage of nuclear waste remains a constraint on any such expansion. The history of the search for a solution to that problem has useful lessons for those considering contemporary social constraints on storage of carbon dioxide and is a matter to which we shall return in the next section (5.2).

Wedges Ten to Thirteen, renewable electricity and fuels, have implications for the marketing of fossil fuels and are all potential or actual areas of investment by oil companies. Indeed, an historian of the oil industry might well study the whole Princeton list and conclude that one company or another had at some stage

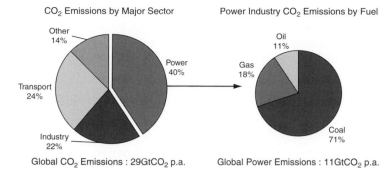

Figure 5.3 Emissions of carbon dioxide by major sector (left) and by fuel type in the power industry (right). The power sector is responsible for about 40% of global emissions and coal represents about 70% of emissions from the power sector: the opportunity for carbon capture and storage from many large stationary point sources is obvious. (A 1-gigawatt coal plant emits 5–7 million tonnes of carbon dioxide a year.) Slide from presentation by Mr Gardiner Hill of BP at ING Oils Forum in London, 2 September 2008 (Kenney, 2008).

invested significantly in those activities. Given that there has been involvement in nuclear power, minerals, coal, even pet food, that claim would be technically correct. But these excursions have tended to be experiments at diversification in good times, to find a home for spare profits, and have been discontinued at times of lower oil price. Will that be the fate of contemporary investments by the industry in biomass fuel, wind power and solar power? The outcome must ultimately depend on regulation and price, yet to be defined.

Substantial research partnerships between universities and oil companies are now examining the underlying science before financial commitments on a classical oil-industry scale are made in areas that may help meet the Princeton targets, such as bio-fuels. In 2007 BP announced a $500 million ten-year partnership with the University of California at Berkeley, the University of Illinois and the Lawrence Berkeley National Laboratory in California, to form an Energy Biosciences Institute. BP's then Chief Scientist, now Undersecretary of Science at the US Department of Energy, Dr Steven Koonin, describes the partnership in BP's *Frontier* magazine in August 2008:

> The Energy Biosciences Institute is operating at the early-stage evaluation of science – in terms of modern biology's contribution to energy – and we therefore needed an open collaborative partnership which would have the ability to explore and be experimental.

By definition the outcome of this BP-style of investment in bioscience research remains uncertain, but it may go well beyond development of effective low-carbon biofuels. What can be said at this stage is that were the oil companies to undertake any large commercial ventures based on such research, these would most probably draw on technologies not traditional to the industry. Not that this is likely to daunt organisations that have proved their ability over the years to apply discoveries in fundamental science effectively at all stages in the oil-industry chain of value, from exploration in frontier areas to manufacture of useful products from hydrocarbons thereby discovered.

Wedges Five, Six, Seven and Eight do draw directly on the distinctive technical skills of the classical oil company. Wedge Five, substituting gas base-load power for coal base-load power involves the exploration and production arms of the oil companies. Wedges Six, Seven and Eight all have subsurface storage at their heart: capturing carbon dioxide at base-load power plants, capturing carbon dioxide at hydrogen plants and capturing carbon dioxide at plants turning coal into liquid fuel (synfuel). Implementation of just one of those three wedges would require geological storage on a massive scale. This storage of carbon dioxide represents an obvious avenue for continuing major involvement of the oil industry beyond its current role, drawing on historically distinctive specialist petroleum engineering and geology – the 'subsurface' technologies.

5.2 SOCIAL SCIENCE DRIVES THE PRINCETON WEDGES

On 12 and 13 October 2005 a conference was held at Burlington House, Piccadilly, London, on *Challenges and Solutions: UK Energy to 2050*. The results of that first meeting were collected and presented to an august group at the Royal Society at a second meeting on 10 November 2005. The weight of the double conference can be judged from the composition of the expert panel on 10 November. In the Chair was Lord Oxburgh, former Chairman of Shell Transport and Trading (Figure 2.4). The other members were Lord Broers, President of the Royal Society of Engineering, Dr Vincent Cable, Liberal Democrat Shadow Chancellor (subsequently hailed in the UK as a prescient hero of the 2008 financial collapse), Hon. Bernard Jenkin, Conservative Shadow Energy Minister and Sir John Lawton, Chair of the Royal Commission on Environmental Pollution (astringent summariser of the BP–ExxonMobil debate in 2003 as then Chief Executive of the Natural Environment Research Council). The spokesman at the crowded main press conference was Dr John Loughead, Executive Director of the UK Energy Research Centre. To the annoyance of Jeremy Leggett (see Chapter 1), chairman of a session on renewables that had taken place on 13 October, much of the media interest was in the suggestion that the nuclear industry would have to play a significant part in providing low-carbon energy to the UK in the period to 2050.

This two-part *Challenges and Solutions* meeting was unusual in several respects. One was the gathering of so many tribes, each with a particular professional stake in the issues discussed and each approaching the discussion from a different angle. The meeting was convened by the principal professional and learned societies with a claim to be closely involved with the topic. These organisations were the Energy Institute, the Geological Society, the Institution of Civil Engineers, the Institution of Electrical Engineers, the Institute of Physics, the Royal Society of Chemistry and the UK Energy Research Council. The conference was supported by the UK Research Councils: Natural Environment, Engineering and Physical Sciences, Biotechnology and Biological Sciences, and Economic and Social. The involvement of the social scientists provided a particularly refreshing new angle for many of the participants, who had long been steeped, even isolated, in the details of the technological arguments.

Sessions on demand, nuclear power, and fossil fuels occupied the first day of the conference. Discussion on renewables and a session on impact took place on the second day. The final invited speaker on day one was Robert Socolow, who took as his title 'Emission free options for energy from fossil fuels: the issues for carbon dioxide sequestration'. He concentrated on one aspect of the Princeton wedges:

> Capturing most of the carbon dioxide emissions from coal and natural gas power plants and storing the captured carbon dioxide in geological formations can become, if deployed on a very large scale within a few decades, one of a handful of principal strategies for mitigating climate change. Early action and learning can be built on low-cost opportunities: 1) capture from a high concentration streams created where natural gas is purified (prior to liquefaction or grid insertion) or where hydrogen is produced by a gasification of coal or [reforming of] natural gas; 2) storage in connection with enhanced oil recovery.
> Carbon policy must elicit early experience with full-scale, plant-lifetime capture and storage at large coal plants, where three billion barrels (one-half cubic kilometre, or 350 billion tons) of supercritical carbon dioxide must be injected over 60 years, generally into non-hydrocarbon brines at depths below 1 km. Policy should also be anticipated that elicits carbon dioxide capture and storage in association with synfuels production.

Socolow offers a constructive alternative to the bleak view of coal attributed to Dr James Hansen, of the NASA/Goddard Institute for Space Studies in New York, in *The Observer* newspaper (18 January 2009): 'Coal is responsible for as much atmospheric carbon dioxide as other fossil fuels combined and it still has far greater reserves. We must stop using it.' Even if we accept Hansen's well-publicised view that we have already driven the concentration of carbon dioxide in Earth's atmosphere to dangerously high levels, we still have to manage the transition to low-carbon sources of energy. As discussed in Chapter 3, during that transition we shall have to continue to depend on coal, oil and gas. Given the very abundance of coal to which Hansen refers, surely it is worth our while to look urgently but carefully at the feasibility of generating electricity from coal while capturing the carbon dioxide thereby released.

The technological demands of carbon capture and storage that might be met by the oil industry are considered in more detail in Chapter 6. What about the demands on society as a whole? How can an energy policy be established within which such activity can take place? This was considered by Professor Susan Owens of the Geography Department at Cambridge University in the final talk on the second day: 'Addressing complex social, cultural and political issues that are key to the viability of a sustainable energy policy'.

I was one of those helping Richard Hardman (see Chapter 3) and his organising group with this meeting. I had urged strongly the inclusion of social scientists in the list of invited speakers. This ecumenical approach had met with initial resistance from some of our colleagues, so I watched with interest while Dr Brenda Boardman of the Environmental Change Institute (formerly the Environmental Change Unit), Oxford University, spoke about housing. Her talk was followed by that of Owens:

> Energy is fundamentally a social and political issue. Energy production involves controversial technologies (old and new) that run into trouble in the absence of public consent. Energy consumption is bound not only to economic prosperity but to cultural norms and taken-for-granted aspects of lifestyle. In a model with which we have long been familiar, there is typically claimed to be considerable technical and economic potential – for a given supply technology, for example, or for energy efficiency gains – which we are failing to realise because of the existence of a social 'barriers'. The implication is that such barriers have to be overcome in the interests of our sustainable energy future. The fact that we said this in the 1970s, and we are still saying it now, should give us serious pause for thought about the validity of this way of thinking. Instead, we need to be treating energy as a socio-technical system, in which social, political and cultural considerations are not 'barriers' to be overcome, but intrinsic to the way in which we think about 'potential' and construct our energy policies.

By the time Owens had finished her talk and she and Boardman had fielded questions from an audience dominated by scientists, it was clear that at least some of those initially sceptical of the value of social science in this discussion had now been won over. A senior oil man, possessed of a pronounced technological approach to business, was expressing his appreciation of what was to him a new perspective. The conjunction in the meeting of the technological framework erected by Socolow and others and the social, cultural and political perspective given by Owens and Boardman had been helpful.

Specifically, the subsurface storage of carbon dioxide described by Socolow can with advantage be considered within the intellectual framework set out by Owens. Elsewhere Owens (2004) explores the dynamics of siting controversies, such as the planning of quarries to provide aggregates for national use in the face of local opposition, and their relationship with political and economic priorities. Her views have direct implications for the storage of carbon dioxide. She urges us (2004, p. 105) to move away from the simple categorisation of siting controversies as national need versus local interests:

> The dominant storyline on siting controversies may now be summarised. Major projects 'in the public interest' are controversial because their impacts fall disproportionately on local communities; as a result there are high costs and long delays in realizing the important benefits of such developments. Salvation lies in adopting a clear strategic framework at national level, which will cascade down to the local level where individual projects must be accommodated. Since issues of principle will have been decided, decisions about such projects can then be made more quickly and efficiently. Whilst conflict can never be eliminated, inclusive involvement of local communities will forge consensus on what constitutes a sustainable development, so that important decisions about the use of land can be made in a less adversarial way.

Owens quickly moves on to recognise that: 'The reality, of course, is profoundly different.' In her October 2005 talk at Burlington House she had harked back to the 1970s. She did not discuss in detail the particular example of the nuclear industry, but for several in her audience this would have come quickly to mind. The 1970s was the time when the issue of safe disposal of nuclear waste in Britain came to be recognised by the public as a critical matter in considering the future development of the nuclear power industry. Subsurface disposal was seen as the key to this matter, which has still not been resolved. The technical lessons from nuclear waste disposal for monitoring underground storage of carbon dioxide have been considered by Stenhouse and Savage (2004): some other aspects are examined here.

Nuclear waste is an intensely political and social issue, just as carbon capture and storage is likely to become. The strategy on nuclear waste in the USA has for over 20 years been dominated by the naming in 1987 of Yucca Mountain in Nevada as the sole site for the final disposal of high-level waste. Some $9.5 billion later, scientists have been unable to find justification in the geology of Yucca Mountain for a premature decision originally based mainly on considerations of politics – not geology – in a sparsely populated area. 'This year, with Barack Obama as president and Harry Reid (Democrat, Nevada) as Senate majority leader, the project could well be dealt a final blow' is the editorial judgement of *Nature* on 5 March 2009, p. 8. Geological endorsement must precede decisions on burying any potentially harmful waste.

That endorsement has to follow careful examination of potential sites for disposal, by geologists working in the field with real rocks. That is why in the late 1970s and early 1980s the geological community in Britain found itself blinking in the unaccustomed glare of publicity. The profession, or at least the geologists working for the government on the matter of safe disposal of nuclear waste, took a public battering in Scotland during the summer of 1980. The tone may be judged by the title of a book published by the Scottish Campaign to Resist the Atomic Menace – *Poison in our Hills: The First Enquiry on Atomic Waste Burial* (SCRAM, 1980). The title is rather more ferocious than the text, which provides some useful summaries of events at and surrounding the Ayr inquiry into test drilling into the Loch Doon granite in southwest Scotland.

The proposal considered in 1980 was not that nuclear waste should be stored in the Loch Doon granite, but that tests of this rock's suitability should be carried out at Mullwharchar, one of the Galloway hills. This anticipates the suggestion of Socolow that tests be carried out to enable sound decisions on subsurface storage of carbon dioxide. The parallels go deeper. We already had nuclear waste to deal with in 1980, just as excess carbon dioxide has already been released to the atmosphere now. The UK was a pioneer with nuclear energy, producing waste of a 'complexity ... [that] requires special consideration' (Royal Society, 2006, p. 1). So why should test drilling not take place in southwest Scotland in 1980? To the distress of many of my political and environmental friends I argued publicly in *The Scotsman* newspaper (4 February 1980) that it should: we needed to know what was feasible and what was not. This was a geological perspective, not a personal or social angle: I was also in favour of considering the suitability of salt in the Cheshire Basin and the London Clay beneath areas surrounding the capital.

There is intellectual discomfort involved in this. I was then, and remain now, broadly speaking a disciple of Dr E. F. Schumacher, a proponent of *Small Is Beautiful: A Study of Economics as if People Mattered* (Schumacher, 1973), and not readily impressed by advocates of a society strongly dependent on centralised technology such as nuclear energy. Yet there I was in 1980 arguing that test drilling for disposal of nuclear waste should take place in the Loch Doon granite – and here I am in 2009 promoting the case for subsurface storage of carbon dioxide. There are a couple of issues to do with nuclear waste on which I am unrepentant.

The first is that we do need to use the subsurface to store safely some of the waste we have already produced, let alone that we are continuing to produce, so we had better understand thoroughly the rocks in any potential storage area. The second is that the argument advanced by Lovelock in *The Revenge of Gaia* in favour of nuclear power is based on his reasoned assessment of priorities in the light of the recent evidence of the seriousness of anthropogenic climate change: it is not some kind of betrayal of commitment to a particular cause. He is right that reducing carbon emissions substantially is an urgent matter, although that does not mean that the world has to plump immediately for French-style reliance on electricity generated in nuclear power stations. The beauty of the Princeton wedges is that they give us a framework for thinking about our options quantitatively and inclusively in our various tribes.

That brings us to the heart of what we might term a Socolow–Owens synthesis. We have a problem with carbon dioxide, for which a plausible and potentially significant solution is capture and storage of that gas. That capture and storage is most commonly considered in connection with large-scale coal-fired power stations. More speculatively, capture and storage could even be involved in future geoengineering to deal with carbon dioxide scrubbed directly from the atmosphere (Kunzig, 2008). On the benign assumption that national, if not international, policy encourages development of carbon capture and storage, whatever the source of the carbon dioxide, the problem then becomes more local as the search for a safe site for storage of the gas begins. Yet policy will need to be underpinned by assurances from scientists that the carbon dioxide can be stored safely following capture. The whole process must bear a cost that will not beggar the operators or wreck the economy, so a fair price must be put on the stored carbon. In closing this virtuous circle the oil industry has the chance to play a leading role.

5.3   A SHARE OF PRINCETON WEDGES FOR THE OIL FOLK

What involvement might the oil companies seek in driving home any of the Princeton wedges? The most traditional use of the industry's skills has been the production of fuel for internal combustion engines. This is still an area where considerable growth in demand may be expected. During the 2003 London debate with BP (Chapter 3), Frank Sprow showed the ExxonMobil forecast that the growth in demand for oil itself will be largely for transport in developing countries (Figure 5.4). The world does love motor cars and there is a deal of effort by technologists to respond to that affection (Graham-Rowe, 2008). These improvements include: hybrid cars, driven by better batteries; better fuel injection; improved design; making existing combustion engines more efficient; and use of hydrogen fuel and biofuels.

The search for satisfactory biofuels has a long history. Yergin (1991, p. 194) reports that early in the twentieth century the British government 'had given over two acres in Dorset to the cultivation of Jerusalem artichokes in the hope that this plant could produce alcohol in commercial quantities to be used as automobile fuel'. There is a strong contemporary revival in work on biofuels, with support from the oil industry. If the world wishes to observe more

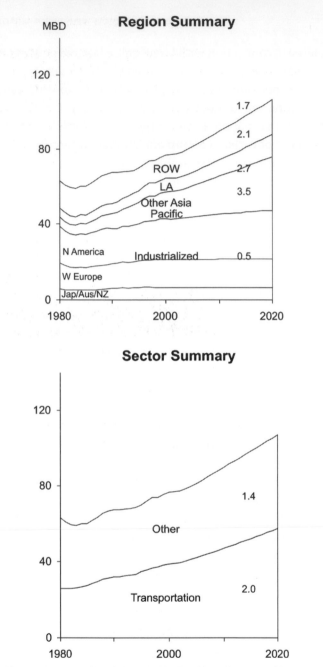

Figure 5.4 World oil demand by region (above) and by sector (below). 'LA' is Latin America, 'ROW' is rest of world. 'MBD' is million barrels per day. Slide shown by Frank Sprow during his initial presentation at the BP–ExxonMobil debate in 2003 (see Chapter 3), to make his point that the growth in demand for oil will be concentrated in developing countries, for use in transport.

thoroughly the moral imperative to feed its people, we shall be ill-advised to take good agricultural land out of food production and use it to fuel personal transport for the rich minority. That imperative is well recognised by responsible opinion: Koonin (2006, p. 435) suggests that 'biofuels could supply some 30% of global demand [for fuels for transport] in an environmentally responsible manner without affecting food production'. That may be, although early experience suggests that the right balance with food production is going to be hard to achieve.

Biofuels could, like oil and gas, be used to generate electricity rather than power cars, trucks, ships and aircraft. Combined with carbon capture and storage this offers a way of not only producing low-carbon energy but even negative-carbon energy. Biomass at large – wood, crop residues and other biological sources – was our first source of energy and until the twentieth century remained the largest. It still comes second only to fossil fuels (Schiermeier, Tollefson, Scully, Witze and Morton, 2008) and remains one of several possibilities for significant low-carbon generation of electricity in this century (Table 5.2).

Electricity generation provides 18 000 terawatt–hours of energy a year, about 40% of our total use of energy (Schiermeier et al., 2008). The main candidates for expansion to make the transition to the low-carbon economy are hydropower, nuclear fission, biomass, wind, geothermal energy, solar, and ocean energy from waves and tides (Table 5.2). The oil industry has so far shown the greatest interest in biomass, wind and solar: in 2009 biomass appears to be favoured at the expense of wind and solar. Only in the case of biomass is there much of a fit with traditional technological skills possessed by the oil folk (those involved in refining), although the scale of engineering and finance involved in the development of wind and solar energy as Princeton wedges is certainly familiar to the major oil companies.

The most familiar part of the Princeton wedges to the oil industry is large-scale development of carbon capture and storage (CCS): CCS is also an important part of the Princeton concept. The storage of carbon provides a particular opportunity for the oil and gas industry to deploy its skills in handling complicated

Table 5.2 *Electricity without carbon, compiled from Schiermeier, Tollefson, Scully, Witze and Morton (2008), who provide some perspective on the numbers in this table. In 2005 (c.9000 hours long), 18 000 terawatt–hours of electricity were generated, giving an average of about 2 terawatts. A gigawatt is 1/1000 of a terawatt. The Hoover Dam on the Colorado River can produce about 1.8 gigawatts. A megawatt is 1/1000 of a gigawatt: a modern train is powered by 3–5 megawatts. A kilowatt powers an electric fan heater.*

Type	Cost ($/kilowatt–hour)	Capacity and comment
Hydropower	0.03–0.10. Installation costs are $1–5+ million per megawatt of capacity; annual operating costs are 0.8–2% of capital costs	Presently 800 gigawatts, c.20% of world's electricity – second only to fossil fuels, and 10 times the combined capacity of geothermal, solar and wind. Could go to c.2400 gigawatts
Nuclear	0.025–0.07, plus large government research and development subsidy	Presently 370 gigawatts, c.15% of world's electricity. At current rate of use and at $130+ per kilogram there is c.80 years' worth of uranium fuel. There are probably large undiscovered deposits of uranium ore: capacity might reach 1000 gigawatts without breeder reactors
Biomass	0.02 burned with coal in conventional plant; 0.03–0.05 in dedicated plant; 0.04–0.9 in co-generation plant with use of waste heat	Presently at least 40 gigawatts, conceivably 5 terawatts by 2050. Biofuels are potentially a more valuable use than electricity generation. Biomass is an important energy source in fires and stoves for more than 2 billion people
Wind	0.05–0.09, competitive with coal at the lower end of the range	Nearly 121 gigawatts at end 2008 (*Nature News*, 12 February 2009): possible 300 gigawatts by 2014. Could grow to 1 terawatt plus with large deployments on plains of China and USA and

Table 5.2 (cont.)

Type	Cost ($/kilowatt–hour)	Capacity and comment
Geothermal	0.05 (best sites)	Presently 10 gigawatts with 3% growth. With a great deal of investment could grow to a terawatt
Solar	0.25–0.40, 0.17 for concentrated solar thermal power. Manufacturing cost of solar cells should fall from current $1.50–2.50 per watt capacity to less than $1 within a few years	In theory the world's entire primary energy needs could be served by less than a tenth the area of the Sahara. In medium to long term the most promising carbon-free technology
Ocean	Tidal 0.20–0.40; wave power up to 0.90	Forty-year-old 240-megawatt plant at Rance, Brittany, remains the world's largest tidal-power installation. Potential European capacity is estimated at 1 gigawatt; global capacity at 40–50 gigawatts (Kowalik, 2004). With a great deal of development wave power might conceivably give 500 gigawatts
Carbon capture and storage (CCS)	Extra costs to reduce carbon emissions from fossil-fuel power stations by 80–90% would add $0.01–0.05 to the cost of a kilowatt–hour. On coal-fired plants technology would be competitive if carbon dioxide were priced at c.$50 per tonne.	Estimates of global aquifer capacity for storage range from 2000 to 11 000 gigatonnes of carbon dioxide. Some 8000 current facilities that might be candidates for CCS emit c.15 gigatonnes annually. The biggest problem is scale: impact within 20 years will require aggressive promotion of the technology

matters below Earth's surface. Oil companies are unlikely to feel the need to seek environmental redemption, especially in the face of uncertain profit. But the carbon-storage business could be profitable for them, given a decent price per tonne of safely buried carbon dioxide. What could the oil industry do to help that come about, to play its part in managing the transition to a low-carbon world while remaining in business?

The answer lies first in considering that question on the very large scale and long term, even beyond the mid-century featured in the concept of the Princeton wedges. Roger Pielke forcefully reminds us (2006, pp. 753–754) that there is a danger of underestimating the challenge of driving carbon emissions still lower in the second half of this century:

> In reality, stabilizing carbon dioxide emissions at current levels, as suggested by Pacala and Socolow, would result in a continued linear increase in atmospheric concentrations because carbon dioxide emissions would still far exceed their rate of removal by the oceans and land. Upon completion of the seemingly herculean task of reducing projected global emissions by more than 50% by 2054, by successfully avoiding seven wedges, we would still face the challenge of reducing the remaining level of emissions by another 64% over the next 50 years. To put their stabilization challenge in stark terms, under Pacala and Socolow's most optimistic assumptions for stabilization at 550 p.p.m., the world will need to reduce its projected business-as-usual emissions by about 1,000 gigatonnes of carbon over the next century. Seven stabilization wedges worth would achieve 175 gigatonnes, leaving a considerable gap, even if the total business-as-usual emissions have been overestimated by a factor of two or more.

Even the oil-reserve optimists agree that by the second half of this century we shall be over the peak of the curve of world oil and gas production from conventional reserves. Unconventional reserves such as oil shale and oil sands, plus coal itself, all carry heavier carbon and other environmental penalties than do conventional oil and gas. So what are those seeking to develop twenty-first-century energy companies to do now? An obvious first step would be to emulate Moody-Stuart at *Coping* with *Climate Change* in 2003, and Tony Hayward at the G8 summit in Berlin in 2007 (see Chapters 3 and 4), and press governments to develop regulatory and economic systems that encourage the development of such long-range solutions as carbon capture and storage.

There is a real urgency about this. Since the concept of Princeton wedges was published in 2004 we have not cut emissions of carbon dioxide, just the opposite. 'News in brief' in *Nature* (2008) provides some numbers:

> Carbon dioxide emissions from fossil fuels and cement manufacturing are rising faster than the worst-case scenario drawn up by the Intergovernmental Panel on Climate Change (IPCC). According to the latest world wide carbon budget, released by the Global Carbon Project, CO_2 levels rose by 3.5% a year between 2002 and 2007, compared with 2.7% as calculated by the IPCC. During the 1990s, emissions rose at 0.9% a year … China is now the biggest emitter of CO_2 and responsible for 21% of the world's emissions – up from 14% in 2002. This knocks the United States into second place, contributing 19% of global emissions. India is fourth, but looks set to take third place from Russia this year.

This rate of increase in emissions of carbon dioxide means that, far from beginning to achieve the Princeton targets, we now require an eighth wedge to meet the mid-century target (Figure 5.5). It may be that yet more wedges are required, to cut emissions even below the levels used in calculating the Princeton wedges. James Hansen (Hansen *et al.*, 2008, p. 218) believes that: 'the present global mean CO_2, 385 ppm [parts per million], is already in the dangerous zone'. A report co-authored by John Holdren of Woods Hole Research Center, Massachusetts (Bierbaum, Holdren, MacCracken, Moss and Raven, 2007) concludes that the atmospheric concentration of carbon dioxide should be stabilised at no more than 450–500 parts per million (ppm), in the lower half of the 450–550 ppm Princeton range.

Professor Holdren has an early opportunity to press his case on cutting emissions where it counts: he took up the post of Science Adviser to President Obama in January 2009. By June he was in Beijing, seeking cooperation between China and the

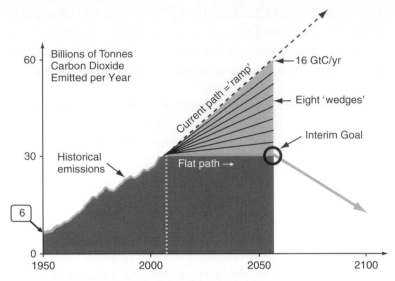

Figure 5.5 A 2008 view of the Princeton 'Stabilization triangle' (see Figure 5.1) shows that eight rather than the seven wedges are now required. Not only have we failed to stabilise our emissions of carbon dioxide since the beginning of this century, the rate of release has increased (see text). Note that '16 Gtc/yr' (16 000 million tonnes of carbon released each year) results from the release of c. 3.7 times that weight of carbon dioxide (vertical axis). BP slide shown at ING Oils Forum (Kenney, 2008).

USA on controlling release of greenhouse gases. The USA still has the highest per-capita emissions of carbon dioxide (see Table 6.2 in the next chapter), even though the USA has now lost its leadership of another dishonourable international league table: the list of the countries with the highest total emissions of carbon dioxide. In that list of total emissions, oil-consuming USA is – as reported by *Nature* (2008) – yielding ground to coal-burning China and India. This changing balance in emissions from oil to coal gives even greater emphasis to CCS. The Princeton targets cannot be achieved without CCS (Figure 5.2) and it is in India and China that the deployment of clean-coal technology is most urgently required. The global role of CCS will be determined mainly in these two countries.

That is on the grand scale. There is another more specific but highly significant opening for the present-day oil companies to play a profitable role in developing a twenty-first-century low-carbon economy: the special skills possessed by their staff. These skills need to be cherished. Although it has been a cliché of business management over the last couple of decades that: 'People are our greatest asset', this has not always been obvious from the actions taken by company executives. The oil industry has followed the crowd in publicly espousing admirable sentiments about staff. It has also regrettably followed the crowd in making redundant chunks of its 'greatest asset' at times of financial stress; redundancies are habitually associated with troughs in the historical cycle of change in the price of oil.

The specialist skills that need to be cherished are, if not unique to the industry, certainly quite unusual. Petroleum engineers and petroleum geologists seek to understand the rocks beneath our feet, how fluids move through those rocks and how those fluids may interact with the minerals lining the pore spaces and pore throats through which they travel. This understanding is just what is required to assess the suitability of any given location for the safe storage of carbon dioxide and to then store that gas securely within the rocks below. It is to these matters that attention is turned in the next chapter.

6

Safe storage: from villain to hero

6.1 ACCEPTING BLEMISHES

Appeals to the conscience of oil folk in relation to climate change are unlikely to have much practical effect. The oil industry can reasonably claim to have performed a useful task that has fuelled much of the prosperity developed over the last hundred years, and it is unlikely to be prepared to take much of the blame for the uneven distribution of that wealth across the world and within nations. Oil company staff are not necessarily disciples of that villain of *Dallas*, J. R. Ewing, although they might aspire to a Southfork-style swimming pool: many have remained aware of their broader responsibilities and have acted on those. The blemishes are of course manifest. Some surface pollution by oil has a natural origin (Figure 6.1), much spillage is emphatically caused by us. It is not possible to deny the Russian imperial legacy to Azerbaijan of oily desolation north of Baku (Figure 6.2); as we have seen, the oil industry did not take any public responsibility for anthropogenic climate change until 1997.

6.2 TAKING RESPONSIBILITY

In his speech at Stanford University in May 1997, announcing the change in policy in BP towards climate change, John Browne took responsibility for reducing the company's operational emissions of greenhouse gases, which amount to a little over 10% of the total volume released through BP's agency. It is relatively normal good practice for a company to take responsibility for its

Figure 6.1 Natural deposit of hydrocarbons at Earth's surface at Pitch Lake, Trinidad. Mr Keith Davison, a consultant with TRACS, is demonstrating the properties of the oil to an (unseen) group of participants in BP's Challenge early development programme. © Claire Budd.

own emissions, and potentially profitable to do so. A notable example of BP acting on that promise to clean up its own operations is the capture and storage of carbon dioxide co-produced with gas in an Algerian operation (Socolow, 2004) (Chapter 4).

More significantly Browne raised the possibility that in time the company could also take a measure of responsibility for the far greater volume of emissions resulting from the use of the company's products by its customers – also the subject of a pledge by ExxonMobil's Frank Sprow in the 2003 London debate with BP (Chapter 3).

In 2005 BP announced plans for low-carbon electricity generation at a plant in northeast Scotland. Carbon dioxide, formed

Figure 6.2 Oily desolation to the north of Baku, Azerbaijan, 2004.
© Claire Budd.

as a waste product in the separation of hydrogen from natural
gas produced from North Sea reservoirs, would be pumped into
the Miller oilfield offshore. The hydrogen would be used to
generate electricity and the carbon dioxide to extend the life of
the Miller field by mixing with and mobilising some of the
remaining oil, thereby enhancing recovery of oil from the
pores of the Miller reservoir. This project needed timely support
from government so that it could begin before the rate of production of oil from Miller, using conventional methods of recovery, fell below a commercial threshold. But the government
decided that all carbon storage schemes had to compete for
funding and tax relief, the winner to be picked after the unavoidable deadline imposed on BP and partners by depletion of the
remaining oil in Miller field. Professor Stuart Haszeldine of
Edinburgh University bemoaned the loss of an 'ideal' project
(*The Observer*, 17 June 2007).

BP subsequently entered a UK government competition,
which was to build a demonstration coal-fired power plant
for carbon capture and storage: distinct from the Miller

gas-to-hydrogen plant discussed above. BP Alternative Energy was on a shortlist of four companies announced in July 2008. But on 8 November 2008 *The Times* reported that: 'The oil company [BP] informed the government last week that it would no longer be submitting a bid for a government-funded scheme to develop a coal-fired power plant using carbon capture and storage … the group had withdrawn because it had struggled to find suitable partners.' Despite that setback in the UK, BP was reported as saying: 'We remain very supportive of alternative energy technology and are we continue to invest in hydrogen energy projects in Abu Dhabi and California.'

International organisations such as BP will choose to invest in countries where government support is tangible rather than notional. Could that in time include the UK? In 2009 *The Observer's* journalists continue their vigorous promotion of the case for the UK to forge ahead with carbon capture and storage (CCS). A news feature on 15 February 2009 explains: 'How Britain can take the lead in the carbon battle', supported by a leading article, 'We must make coal-fired energy less toxic', which links CCS with energy security. This leader claims that 'CCS could yet be a lucrative component in Britain's post-recession economy' and calls for a firm commitment by government backed up by 'substantial financial investment'.

Just a few weeks later (26 April 2009), following positive announcements on CCS in the UK annual budget, *The Observer* was carrying a triumphant leader: 'A victory for green thinking'. The Rt. Hon. Ed Miliband, Secretary of State for Energy and Climate Change, wrote in *The Times* (27 April 2009): 'the most important technology the world can develop is the technology to capture carbon emissions and store them permanently deep underground'. The UK appeared to be taking a lead in developing coal power with CCS, thereby coupling the energy security provided by the abundance of coal with promoting the development of a low-carbon economy.

Whatever the inevitable variations in government support from country to country, the time for the oil industry at large to take some significant responsibility for their customers' behaviour may now have come. Pacala and Socolow (2004), the IPCC

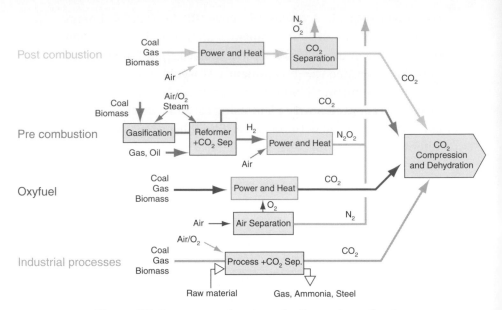

Figure 6.3 Processes and systems for the capture of carbon dioxide at power stations. The fuel may be biomass as well as gas and coal, but the biggest prize appears to be the generation of low-carbon electricity from coal (see text). Figure from Intergovernmental Panel on Climate Change (IPCC) (2005, their Figure TS-3).

(2005), Gibbins *et al.* (2006) and Shackley and Gough (2006) show the large part in reducing emissions that can be played by capture and burial of fossil carbon once it has been used to generate energy. Large coal-fired power plants are prime candidates for application of this technology.

There are several means of capturing carbon dioxide from a fixed point source (Figure 6.3), all of which involve some loss of overall efficiency at the power plant (Figure 6.4). The oil industry would prospectively become involved at the stage of compression and dehydration of the carbon dioxide at the power plant, ready for transport by pipeline and storage in rocks.

That storage could in time involve technological development that led to safe retention of carbon dioxide on the deep-sea floor (Palmer, Keith and Doctor, 2007). But the present emphasis is on subsurface storage, both onshore and offshore. The industry

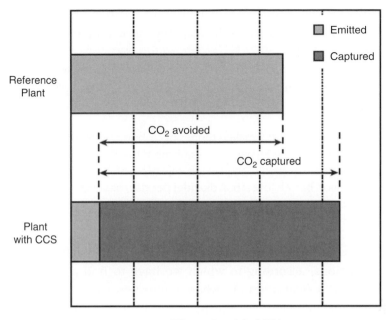

Figure 6.4 Loss of overall efficiency in power plants as a result of capturing carbon dioxide. Horizontal axis is based on kilograms of carbon dioxide produced per kilowatt–hour. Figure from IPCC (2005, their Figure TS-11).

has the experience, the technology and enough access to safe locations, to play a leading role in that underground storage. At the risk of infuriating their opponents still further, the oil companies can enjoy the prospect of being paid twice: once to take the fossil carbon out of the ground and again to put it back.

6.3   OIL RESERVOIRS TO THE RESCUE?

Reservoir geologists and reservoir engineers, working for a range of state and other oil companies, have traditionally published details of their operational application of science. In many cases they have done this in close collaboration with universities and have thereby contributed substantially to the international peer-reviewed literature. New and fundamental insights have arisen from this activity. There is already a large

body of freely accessible information that can guide the underground storage of carbon dioxide (see for example Baines and
Worden, 2004a).

The volumes of carbon dioxide required to make a dent in
the Princeton targets are huge. Schiermeier (2006, p. 621) gives
some figures:

> Locking away 250 million tonnes of carbon [915 million tonnes
> of carbon dioxide] per year – equivalent to 4% of annual global
> emissions – would require an injection of 25 million to 35
> million barrels [of carbon dioxide] per day, depending on com-
> pression density. That's equivalent to about a third of the flow
> of oil per day currently coming from reservoirs.

These volumes have to be set against the original (2004) Princeton
calculations, according to which we have to find a way *not* to
release 175 gigatonnes of carbon (equivalent to 640 gigatonnes
of carbon dioxide) over the next 50 years. Yet there is some
consolation to be found in the oilfields themselves.

The term 'oil industry' is in one sense misleading: the major
product is salty water pumped to the surface with the oil. (Oil-
bearing rocks also contain original water in their pores which
is produced with the oil; oil reservoirs produce more and more
water as the oil is pumped out, to be replaced in the pores by
inflowing water.) Daily oil production of over 80 million barrels
is accompanied by well over 200 million barrels of water (Veil,
Puder, Elcock and Redwell, 2004). Reversing that total flow of
some 300 million barrels a day, by injecting into existing reservoirs (say) 2500 million tonnes of carbon a year, in the form
of over 9000 million tonnes of carbon dioxide, would cope with
125 gigatonnes of carbon (450 gigatonnes of carbon dioxide),
about 70% of the 2004 Princeton target for controlling emissions
through to the middle of the twenty-first century. That is not
going to happen, but it does give some indication of the scale of
the contribution that could be made by the use of the existing
oil reservoirs.

If we consider the potential use of rocks that do not contain
oil and gas, the 'saline aquifers' that contain salty water in their
pores and fractures rather than hydrocarbons, the scope for

storage increases much further. But injecting carbon dioxide into a saline aquifer is quite a different matter from using a reservoir that has previously contained oil and gas and therefore has a proven capacity to retain safely injected carbon dioxide up to a certain pressure. We are some way from understanding well enough the effects of injecting carbon dioxide into saline aquifers to be able to commit with confidence to the widespread use of these aquifers for carbon capture and storage projects. Expertise that already exists in the oil industry can help greatly in developing that understanding.

There are some indications in an arcane branch of reservoir geology that a natural process associated with oil maturation and migration may be turned to good effect in storing carbon dioxide in saline aquifers. Observations of the microscopic details of certain oil reservoirs suggest that a volume of oil migrating from the carbon-rich rocks that are the source of the oil to the porous 'reservoir' rocks that contain the oil accumulation, is preceded by a volume of acidic fluid that etches microscopic holes and channels in the rocks through which it passes (Schmidt and MacDonald, 1979a,b) (Figure 6.5). Comparison of sandstones that contain carbon dioxide-rich fluids and those that do not (Pearce *et al.*, 2004) suggests that carbon dioxide in solution (carbonic acid) can be involved in the creation of this 'secondary porosity'.

It is theoretically possible that an astute combination of reservoir geology and reservoir engineering could both create storage space in a suitable saline aquifer by such dissolution and then also seal that body of rock with an impermeable layer (Johnson, Nitao and Knauss, 2004) at an appropriate and distant location by controlled re-precipitation of the dissolved material. A crucial matter here is the rate at which the reservoir fluids, made more acid by the addition of injected carbon dioxide, react with the reservoir rock itself.

Baines and Worden (2004b) report:

> Examination of a CO_2-filled porous sandstone with abundant reactive aluminosilicate minerals [e.g. clay minerals] that received a huge CO_2 charge about 80 000 to 100 000 years ago reveals minimal evidence of solid phase sequestration of the added CO_2.

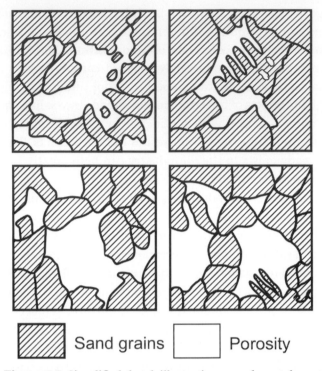

Sand grains ☐ Porosity

Figure 6.5 Simplified sketch illustrating pore-shapes characteristic of secondary porosity in sandstones, after Schmidt and McDonald (1979b, their Figure 7). Secondary pores are formed by dissolution of individual grains of sand and the primary cement that binds them together, following the deposition and burial of a body of sand. They can be detected by microscopic examination of rocks: the very stuff of professional life to reservoir geologists. The origin of these secondary pores is believed to be relevant to present-day storage of carbon dioxide.

Gilfillan *et al.* (2009, p. 614) cite evidence that the trapping of carbon dioxide in pore-water is an order of magnitude greater than trapping of carbon dioxide by reaction with the minerals lining those pores. They add advice for those who would store carbon dioxide in porous rocks:

> In view of our findings that geological mineral fixation is a minor CO_2 trapping mechanism in natural gas fields, we suggest that long-term anthropogenic CO_2 storage models in similar geological systems should focus on the potential mobility of CO_2 dissolved in water.

Yet Houston, Yardley, Smalley and Collins (2007, p. 1146) observe quite rapid rates of reaction between reservoir rocks and seawater injected to increase production from a North Sea oil reservoir:

> Our observations demonstrate that silicate–water reactions [that is, reactions between the injected seawater and the pore-lining minerals] proceed on a time scale of months in the Miller Field, and so such reactions are likely to have the potential to provide an effective means of fixing CO_2 as alkali bicarbonate on a similar time scale.

Houston and her colleagues (2007, p. 1143) draw a significant general conclusion from their North Sea study:

> The safety case for CO_2 storage in such reservoirs is greatly facilitated if it can be shown to react with the host pore waters and rocks on a human time scale, and the results of this study indicate that this is indeed the case.

What are we to make of these apparently discrepant results from Baines, Worden, Gilfillan and colleagues on the one hand, and Pearce, Houston and colleagues on the other? Just as 'secondary porosity' (Figure 6.5) developed extensively only in particular geological settings in the past, so we may today expect to see considerable variations in reactivity of carbon dioxide with pore-lining minerals across a range of prospective storage sites. Our approach to storage will be modified as the results from CCS pilot studies become available.

Armed with the results of those CCS pilot studies, we can maximise both the storage capacity of reservoirs, and the safe retention of injected carbon dioxide, by combining geology and engineering in a cross-tribal fashion. Also, if carbon dioxide is to be injected into existing oil and gas fields, reactions of carbon dioxide with the cement and steel components used to construct and control the wells need to be examined carefully (Rochelle, Czernichowski-Lauriol and Milodowski, 2004). This would be standard work in a modern oil company, with staff trained to think and work across the boundaries of the classical geological and engineering departments.

The various chemical reactions discussed above would take place within a contained reservoir well below the ground, with an effective seal of impermeable rock preventing escape of the carbon dioxide back to the surface. A clear distinction should be drawn between this method for storing carbon dioxide and that proposed for subsurface disposal without the existence of an effective seal provided by layers of overlying impermeable rocks. Kelemen and Matter (2008) have proposed injection of carbon dioxide into a particular type of igneous rock – peridotite – with which it would react rapidly. In this case the trapping mechanism would be the formation of carbonate by a reaction between the injected carbon dioxide and the peridotite. Peridotite is formed at depth, in the mantle, but may be exposed at Earth's surface as the result of uplift associated with the movement of tectonic plates.

Kelemen and Matter have studied the rate of natural carbonation of mantle peridotite uplifted and exposed in the mountains of Oman and have concluded that this process could be replicated by man to trap more than a gigatonne of carbon dioxide a year in Oman alone. If this were to prove feasible, it would clearly make a significant contribution to controlling carbon emissions – maybe as much as half a Princeton wedge (Chapter 5). Particular drilling and injection techniques long familiar to the oil industry would play a central part in any subsurface storage of carbon dioxide in peridotites in Oman or elsewhere. An obvious concern would be how to control, in the absence of natural seals, potentially dangerous escape of any untrapped carbon dioxide back to the surface. Given that safety is a priority, pilot ventures in subsurface storage are likely to remain focussed on sealed reservoirs.

Assessment of subsurface chemical reactions with injected carbon dioxide provides just one example of how the expertise of the oil industry, built up during a century of collaboration with the world's universities, might be used to cope with the problems created by use of the industry's main product. Whether these techniques are properly evaluated at all, let alone in time to make a difference, is only partly the responsibility of the oil companies themselves. The technology is advancing rapidly, but governments are moving rather more slowly (Stephens and Van Der Zwaan, 2005). There has to be a framework of government regulation and policy to create the economic circumstances in which the necessary activity can take place (Nicholls, 2007b).

6.4 WIDESPREAD UNDERGROUND STORAGE

On the largest scale geologists use their knowledge of the workings of the whole planet to predict the location of possible hydrocarbon reservoirs within a sedimentary basin. Like the variety of basins in which we wash our hands across the world, geological basins come in a range of shapes and sizes. They may extend over thousands of kilometres. (The example of locating the Forties field in the North Sea basin is considered in Chapter 8.) In frontier exploration of new basins, oil companies will seek to negotiate rights over large areas. A joint venture between BP and the Turkish State oil company TPAO that began in the 1990s covered the entire southeast region of the deepwater Black Sea – an area of some 76 000 square kilometres.

The search for suitable rocks in which to store carbon dioxide will begin on a comparably large scale (Chadwick, Arts, Bernstone, May, Thibeau and Zweigel, 2008). Locating existing oil and gas reservoirs is simply a matter of looking at maps that are freely available. Finding other bodies of rock that do not contain hydrocarbons but which might still be suitable for storing carbon dioxide is a different matter. That leads to a series of important distinctions.

If carbon dioxide is to be stored only in oil reservoirs in which there is a prospect of thereby recovering additional reserves ('enhanced recovery'), storage capacity will be limited. If oil and gas reservoirs in general are to be used, without the additional bonus of extending the life of the field, that increases storage capacity considerably. Then if we add porous rocks that are not oil and gas reservoirs, the 'saline aquifers', storage capacity increases by an order of magnitude – but certainty of safe storage decreases. UK North Sea oilfields have an estimated storage capacity, when enhanced oil recovery is involved, of the order of 700 million tonnes of carbon dioxide. Yet the total capacity of all the oil and gas fields and aquifers in the UK sector of the North Sea is estimated to be 20 000 – 260 000 million tonnes of carbon dioxide

(Mather, 2005): a large range that reflects the uncertainty in the early stages of this assessment of capacity for storage in saline aquifers.

These distinctions are clearly important in considering total storage capacity. Another large-scale factor is how convenient these bodies of rock may be for disposal of carbon dioxide from, for example, large power stations located onshore near to sources of domestic or imported coal. The costs of transport of carbon dioxide are low compared with costs of capture (see discussion later in this chapter), so pumping carbon dioxide to remote locations may be feasible. This is just as well given the potential controversy discussed in Chapter 5: potential disposal sites are likely to arouse concern in society comparable to that caused by the debate on disposal of nuclear waste.

Consideration of these large-scale elements does support the view of Haszeldine that the Miller project was indeed an ideal way to begin the UK programme of carbon capture and storage. It would partly pay its way by recovery of more oil. Hardcore environmentalists might well splutter at this stage, but on a broader environmental view, it is sensible to squeeze the last drop of oil out of existing fields in the excellent cause of developing a central plank in a low-carbon economy. Miller field is offshore, far from any potentially nervous denizens of the UK, and it has a proven capacity to trap fluids safely over millions of years.

That does not mean such a project is without risk. Any such venture needs to be analysed systematically on a whole range of scales, from thousands of kilometres across a sedimentary basin to a few microns across the narrow connecting channels between pores in an oil reservoir. Chadwick *et al.* (2008) provide a handy list of things to be checked before use of a saline aquifer for storage of carbon dioxide. The use of a saline aquifer is the most demanding case, because at the start there cannot be the reassurance offered by the proven capacity of a hydrocarbon reservoir to retain fluids over long periods of time.

The first stage in considering a saline aquifer is site screening, ranking and selection, in which geology has a key role (Table 6.1). The selection process includes consideration of storage capacity, basic properties of the reservoir, basic properties

Table 6.1 *Key geological indicators for suitability of a saline aquifer for storage of carbon dioxide. Based upon Table 3.1 in Chadwick, Arts, Bernstone, May, Thibeau and Zweigel (2008) with permission of the British Geological Survey.*

	Positive indicators	Cautionary indicators
Reservoir efficacy		
Static storage capacity	Estimated effective storage capacity much larger than total amount of carbon dioxide to be injected	Estimated effective storage capacity similar to total amount of carbon dioxide to be injected
Dynamic storage capacity	Predicted injection-induced pressures well below levels likely to induce geomechanical damage to reservoir or caprock	Injection-induced pressures approach geomechanical instability limits
Reservoir properties		
Depth	Over 1000 metres, less than 2500 metres	Less than 800 metres, greater than 2500 metres
Reservoir thickness (net)	Greater than 50 metres	Less than 20 metres
Porosity	Greater than 20%	Less than 10%
Permeability	Greater than 500 millidarcys	Less than 200 millidarcys
Salinity	Greater than 100 grams per litre	Less than 30 grams per litre
Stratigraphic	Uniform	Complex lateral variations and complex connectivity of reservoir facies
Caprock efficacy		
Natural continuity	Stratigraphy uniform, small or no faults	Lateral variations, medium to large faults

Table 6.1 (*cont.*)

	Positive indicators	Cautionary indicators
Thickness	Greater than 100 metres	Less than 20 metres
Capillary entry pressure	Much greater than the maximum predicted injection-induced pressure increase	Similar to maximum or predicted injection-induced pressure increase

of the overburden and basic simulations of flow in the reservoir. Then safety must be assessed, as must conflicts of use and costs.

Once a site is selected it has to be characterised. The first step is to describe the geology, followed by predictive modelling of flow. Then come geochemical and geomechanical assessments, leading to formal assessment of risk. The monitoring programme has to be designed and the engineering of the well has to be given close attention. If carbon dioxide is to be stored safely, the integrity of the wells has to be maintained over the operational life of the storage site and afterwards. Then consideration can be given to the method of transport of carbon dioxide to the site.

This is followed by design of the site and securing national and international planning consent. Only then can consideration be given to site construction, the operations themselves and closure. At this stage there continue to be crucial issues: criteria for safe site closure, clarity about who owns the pore space, the transfer of liability from operator to national authority and the huge question of the estimated retention time of carbon dioxide within the reservoir and its ultimate fate.

A similar type of list could be constructed for the life of an exploration and production venture in the oil industry (Jahn, Cook and Graham, 2008). A powerful educational tool used in basic industry training programmes is the concept of the 'chain of value' leading from gaining access to prospective acreage in a sedimentary basin, assessment of that licensed area by mapping, remote sensing by seismic techniques, followed by exploratory drilling. If that drilling finds oil, it is followed by appraisal of the size and quality of the reservoir. If oil and gas are present in commercial quantities, the chain of value leads on to the design of facilities and their construction. Then production begins, climbs to a plateau, falls away to the point where money is no longer to be made, followed by safe abandonment. In an extreme case, as with the onshore Wytch Farm oilfield in southern England, a condition of the licence could be that every single piece of equipment would be completely removed and the land restored to its original condition.

In the case of the oil industry 'chain of value', there is no income until production begins, by which stage hundreds of millions if not billions of dollars may have been committed to the project. Worse, it is a truism that the size of an oil or gas field is never known until the last barrel of oil or cubic foot of gas has been produced. Expense and uncertainty are combined in an unsettling manner.

Useful parallels can be drawn between operations in the oil-industry chain of value and the subsurface storage of carbon dioxide. The most obvious of these parallels is that at the centre of the business is an understanding of the reservoir itself and the fluids it contains. This understanding evolves as time goes by because conditions within the reservoir are changing as the operation proceeds, whether that operation is to remove oil or to store carbon dioxide. There are obvious changes in the volume of oil and gas with changes in pressure, and the same is true of supercritical carbon dioxide (Figure 6.6). There will also be changes in composition of the mixture of fluids as time goes by. Some of these changes will be the result of chemical reactions taking place between the fluids the minerals lining the pores. These reactions cause changes in the geometry of those pore spaces and may also affect the properties of adjacent rocks, including the sealing capacity of the originally impermeable rock lying above the reservoir.

So it is clear that petroleum geologists and petroleum engineers have a flying start when it comes to the business of storing fluids underground. They have encountered natural accumulations of carbon dioxide in their search for oil and gas and even

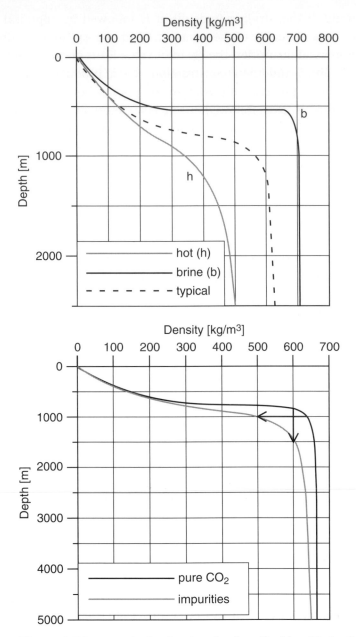

Figure 6.6 Increase in the density of carbon dioxide with depth of burial. The vertical axis shows the depth in metres, the horizontal axis density in kilograms per cubic metre. In the typical case carbon dioxide becomes 'supercritical' at a depth of around 800

injected carbon dioxide to enhance recovery of oil. They would prefer to discover methane rather than the natural accumulations of carbon dioxide documented by Pearce *et al.* (2004) but would have no doubt that carbon dioxide may be safely trapped in reservoirs over geological timescales.

Remote sensing using 'seismic' techniques has for decades been a key tool for those looking for oil and gas (see Figure 8.4 for an example and an explanation). More recently repeated seismic surveys have been used to monitor depletion of oil and gas reservoirs. This technique is now used to assess the growth of accumulations of carbon dioxide (Figure 6.7) being pumped into the Utsira Sand reservoir at Sleipner in the North Sea, as part of a Norwegian CCS operation (Chapter 4). These seismic data enable modelling of the flow, leading to an assessment of the implications for underground carbon storage (Bickle, Chadwick, Huppert, Hallworth and Lyle, 2007). Modelling of injection of carbon dioxide into well-studied mature fields such as Forties can be attempted even without direct observation of the process (Ketzer, Carpentier, Le Gallo and Le Thiez, 2005).

So far, so good; there is a feeling that we are in familiar territory and that the technical issues to do with fluids, rocks and pipes can be thought through by experts using a sensible framework such as that set out by Chadwick *et al.* (2008), leading to a good degree of confidence in the outcome of the operations themselves. For the record, it should be noted that Drs Andy Chadwick, Sam Holloway, Nick Riley and their colleagues in the British Geological Survey (BGS) have played a pioneering role in tackling these technical matters, without an initially warm response from others. I recall getting a cool reaction from some academic and industry colleagues when I would return from Board meetings of BGS in the 1990s, bouncing like Tigger with

---

Figure 6.6 (cont.) metres: below that depth it will expand to fill pore space like a gas but will have a density comparable to that of a liquid. The 'hot' case is with an elevated geothermal gradient of 45°C per kilometre. The 'impurities' case is with 2.75% oxygen and other components. Based upon Figure 3.2 of Chadwick, Arts, Bernstone, May, Thibeau and Zweigel (2008) with the permission of Vattenfall and the British Geological Survey.

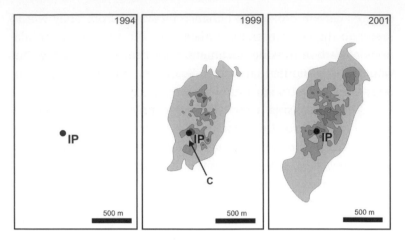

Figure 6.7 The progressive development from 1994 to 2002 of a plume of carbon dioxide stored in the Utsira Sand 'saline aquifer' up to 1 kilometre beneath the bed of the North Sea at Sleipner, showing the developing plume in plan view. This style of monitoring of the annual injection of c. 1 million tonnes of carbon dioxide is made possible by the use of remote sensing techniques ('time-lapse seismic') developed by the oil industry to measure the rate of depletion of oil and gas reservoirs (see Figure 8.4 for an example and explanation of seismic data). The carbon dioxide is produced in association with hydrocarbons. Its capture and storage make commercial sense because of regulations introduced by the government of Norway. Point of injection is marked 'IP'; the impressive detailed interpretation of the position of a 'primary feeder chimney' is shown by 'C'. Scale is 500 metres. Based upon Figure 7.2 of Chadwick, Arts, Bernstone, May, Thibeau and Zweigel (2008) with the permission of the British Geological Survey.

enthusiasm for the research on the storage of carbon dioxide that I had seen at the Survey's laboratories at Keyworth, England. Now at least some of the early doubters have become active zealots. It is becoming clearer that given a suitable regulatory and economic framework, this could prove to be a decent business to be in.

But when we come to planning consent and issues of liability, we enter the difficult territory of perception of risk by those who are not necessarily experts. Experience with the issue of disposal of nuclear waste is not an encouraging precedent, although we do now have the Socolow–Owens intellectual framework described in Chapter 5 to guide us. Can the risk of escape of carbon dioxide from underground storage be minimised to the point where doubters are reassured?

The route to widespread carbon capture and storage that is the most reassuring from a safety angle, and also partly self-funding, is to pump carbon dioxide through a pipeline from a power station into existing oil reservoirs to enhance recovery of the remaining oil. Eventually a safe quantity of carbon dioxide will be left in the reservoir and sealed off with every reason to expect it to remain safely underground on geological time-scales – especially in the common case of an overlying mudstone seal (Lu, Wilkinson, Haszeldine and Fallick, 2009). The problem is that the amount of carbon dioxide that can be conveniently stored in existing oil reservoirs is small compared with the quantities required to satisfy even one of the Princeton wedges discussed in Chapter 5. Tackling the big issue will require the use of saline aquifers.

6.5   COAL POWER WITH CARBON CAPTURE AND STORAGE

Although there has been a focus on styles of personal transport in the climate change debate in Europe and North America, the numbers indicate clearly that the main issue lies in Delhi and Beijing, not in Detroit and Berlin (Table 6.2) (Liu and Diamond, 2005). Can developing nations such as Vietnam (Figure 6.8) achieve their ambitions for cars and prosperity without emulating twentieth-century energy practice in developed countries? Specifically, can carbon dioxide captured from coal-burning power stations in China and India be stored safely underground in those countries? The short answer is that this cannot be done using just existing oil and gas reservoirs, because they do not have sufficient capacity in the locations required. Storage of the huge volumes that would make a significant difference globally would entail the use of rocks that have not previously had oil and gas drained from them, the saline aquifers. Before these can be used with confidence we shall have to build up a detailed understanding of the processes involved, not least by experience with refilling spent oil reservoirs with carbon dioxide.

Table 6.2 *Carbon dioxide outputs for China and selected other countries.*
Data from Liu and Diamond (2005).

Country	Population in millions in 2003	Carbon dioxide emission in tonnes per capita in 2000	Total carbon dioxide emission in million tonnes in 2000
China	1288	2.2	2780
India	1064	1.1	1120
Japan	127	9.3	1180
Russia	143	9.9	1440
USA	291	19.8	5590
World	6271	4.0	24210

Figure 6.8 Ho Chi Minh City, Vietnam, October 2003. Motor
cycles had by this time already replaced many bicycles since my
first visit to the city in September 1997. How long before cars
dominate the street? Photograph by my colleague Claire Budd from
her vantage point on the front seat of our official BP transport
(not a bicycle). © Claire Budd.

The prize for being the first town to have a coal-fired power station that buries its own carbon dioxide may be claimed by Spremberg, Germany. *The Observer* (28 October 2008) describes the transition:

> It used to be called Stinky Town – a smoke-belching, coal-burning industrial powerhouse in what was once the heart of East Germany … It is now host to the world's first power plant that collects emissions from coal burning and pipes them deep underground. Built by Swedish power firm Vattenfall, it emits up to 90% less carbon dioxide than a conventional facility. The process it employs, known as carbon capture and storage (CCS), is seen by a growing coalition of power companies, financiers, academics – and even campaign groups including Greenpeace and WWF – as the silver bullet in the fight against climate change.

Chadwick *et al.* (2008, p. 6) report that '[a] case study has identified and characterised deep saline aquifers of regional extent in the North-east German Basin with a number of potential storage sites' for the captured carbon dioxide, which would otherwise be an invisible component of the more obvious pollution at 'Stink-Stadt' (*The Observer's* term). Vattenfall is developing a 30-megawatt pilot-scale power plant, with facilities to capture carbon dioxide, on the same site as an existing lignite-fired power station at Schwarze Pumpe, Spremberg. Chadwick *et al.* (2008, p. 6) give some perspective on the numbers involved:

> Amounts of this gas [carbon dioxide] produced from the pilot plant will, at full-load operation, be about 60 Kt per year, compared to the 10 Mt per year produced currently at Schwarze Pumpe. CO_2 volumes that will need to be handled from high-range point sources are large, with a project scale significantly bigger than any of the current worldwide CO_2-storage projects that are in operation today. For example, the annual CO_2 amounts stored at Sleipner, In Salah (Algeria), and Weyburn (Canada) are in the range about 1 to about 2 Mt.

We may compare these figures with the gigatonnes involved in the Princeton wedges discussed in Chapter 5. The economic incentive for action has to be clear before we can move to the scale of carbon capture and storage required, which is far beyond the pilot study in Germany. How might that be achieved in Europe?

The BBC's environment correspondent Mr Roger Harrabin (2008) sets out some options:

> New rules mandating that all coal power plants must be fitted with CCS – i.e. industry and the consumer will pay.
> Direct funding from the EU or member states (but member states do not want to pay).
> A feed-in tariff so generators get a premium for the amount of CO_2 they sequester. Monitoring may be difficult.
> Creating a new fund within the EU's Emission Trading Scheme (EUETS) which would give firms valuable carbon credits for every stored tonne of CO_2. This would cost nothing but might undermine the CO_2 market.
> Setting a CO_2 emission limit for all new power stations of, say, 350g of CO_2 per kilowatt hour of electricity. This would make it impossible to build a coal plant which did not capture at least as some of its CO_2. This option is being pushed hard by UK Conservatives.

Harrabin's summary gives some indication of the complications and ramifications that will accompany the introduction of any regulatory framework to achieve the overriding objective of encouraging the development of low-carbon generation of electricity from coal, in Europe or elsewhere. The first option listed by Harrabin, that all coal-fired power plants be fitted with carbon capture and storage was an early favourite, following a decision of a European Parliament committee reported in *The Times* on 8 October 2008. But only days later (17 October) the same paper had the headline 'Black clouds hanging over green targets as EU states say we can't afford them'. The global financial crisis was testing thoroughly the resolve of the legislators: several had lost their environmental nerve. The collapse to below 20 euros in the price of the right to emit 1 tonne of carbon dioxide to the atmosphere reflected the fall in the wider commodity markets and did nothing to help faltering governments sustain their professed faith in green policies.

Those days in October 2008 epitomised the cliché that a week is a long time in politics. Those involved with the oil industry have more opportunity than most to understand that a week in the life of planet Earth, barring asteroids and earthquakes, is generally pretty small beer. Armed with that perspective, can the oil companies help the legislators to keep their nerve? Is there an industry approach to low-carbon economics that will prove robust through financial turmoil?

6.6 ENHANCING THE ECONOMICS

Injection of carbon dioxide, discussed earlier in this chapter in relation to the aborted Miller project in the UK North Sea, is already common practice in the industry, especially, and somewhat ironically, in Texas (Figure 6.9). Injection is a means for enhancing recovery of residual oil in reservoirs from which much of the oil has already been extracted. That prospect of enhanced oil recovery is the most likely opening for the oil industry

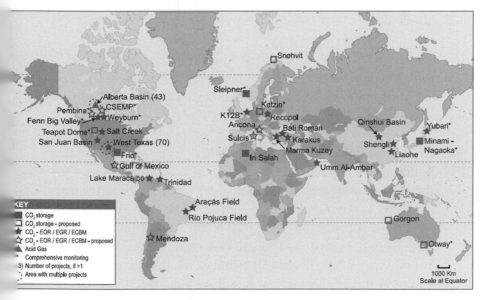

Figure 6.9 Sites of injection of carbon dioxide. Note the concentration in Texas and Alberta. EOR = enhanced oil recovery; EGR = enhanced gas recovery; ECBM = enhanced coal bed methane. There is industrial-scale (~1 million tonnes a year) injection of carbon dioxide at three locations: Sleipner in the North Sea, Weyburn in Canada and In Salah in Algeria. From IPCC (2005, their Figure 5.1).

Figure 6.10 Hardworking and quite muddy team of BP reservoir geologists, reservoir engineers and sedimentologists, recovering core of Forties Formation reservoir on the deck of Forties Delta on 17 May 1982. From Lovell (2008), photograph by the author.

to enter decisively into this new business of storing carbon dioxide – and recovering extra oil has long been an obvious industry objective.

At a time of record oil prices in the early 1980s, BP ran an experiment in 'enhanced oil recovery' using surfactants in Forties field (Figure 6.10, Figure 6.11). BP reservoir geologist Dr Chris Sladen (now President, BP Mexico) led the work on the fine details of reservoir geology, especially the post-depositional changes in porosity following burial. For the technical record, these include beautifully formed new crystal faces of the mineral quartz in places (Figure 6.12) and development of book-like structures of the clay mineral kaolinite (Figure 6.13). Other oil companies and universities were engaged in comparable studies and wanted to exchange information and ideas. So, in August 1982, Sladen and I flew first-class (yes, the oil price was *really* high!) to Hawaii to present the Forties data at a Geological Society of America Penrose Conference on diagenesis (changes in sedimentary rocks following burial). Among those we met there with

Figure 6.11 BP's Chief Sedimentologist pictured relaxing following core recovery on the deck of Forties Delta on 17 May 1982. Note the immaculate and mud-free protective clothing. From Lovell (2008), photograph by an admiring colleague.

common interests was Dr Yousif Kharaka of the United States Geological Survey.

Now, over a quarter of a century later, Kharaka comes back into this story. A recent paper of which he is senior author (Kharaka *et al.*, 2006) concerns the effects on the rocks themselves of storing greenhouse gases in sedimentary basins and lies on my desk as I write this. Kharaka is still looking at reservoirs, but now with a view to putting fossil carbon *back*. There are plenty of prospective areas in which this might be done (Figure 6.14). The

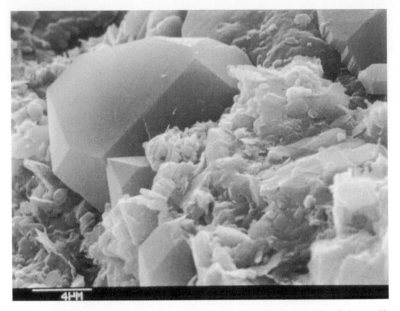

Figure 6.12 Forties Sandstone reservoir: photograph of the wall of an individual pore taken by Chris Sladen, using a scanning electron microscope. It shows the crystal faces of overgrowths of quartz on an original grain of sand. The scale bar is 4 microns (4 millionths of a metre).

oil business will consider favourably the idea of storing carbon if it can make a profit commensurate with the risks. As well as considering the future price of a barrel of oil, the prospective value of a tonne of carbon dioxide put safely back underground becomes crucial.

The economic case was summarised by Mr Gardiner Hill and Mr Bob Edwards at an ING Oils Forum in London on 2 September 2008 (Kenney, 2008). Hill is Director of Carbon Capture and Storage Technology at BP; he is also the person who works with Princeton on the Carbon Mitigation Initiative – including the 'Wedge Project' (Chapter 5). Edwards is BP Vice President Coal Conversion, CCS and CO_2 Management. Hill and Edwards considered the feasibility and cost of solving the greater part of a Princeton wedge with carbon capture and storage. If half of all new coal-fired power plants were to be built to capture

Figure 6.13 Forties Sandstone reservoir: kaolinite has grown across the throat of a pore in the form of pages in a microscopic book. Photograph by Chris Sladen, using a scanning electron microscope. The field of view is c.3 microns (3 millionths of a metre).

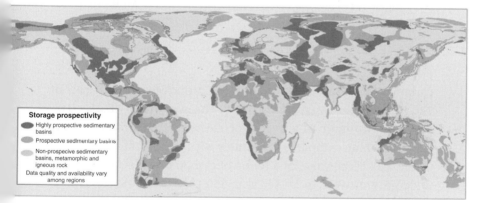

Storage prospectivity

- Highly prospective sedimentary basins
- Prospective sedimentary basins
- Non-prospecive sedimentary basins, metamorphic and igneous rock

Data quality and availability vary among regions

Figure 6.14 Suitability of sedimentary basins for subsurface storage of carbon dioxide. Basins containing sedimentary rocks come in a range of sizes and shapes around the world, just as – on a rather smaller scale – do basins and kitchen sinks in different countries. Potentially useful layers of sedimentary rocks may still have some remaining oil and gas in their pores, or these pores may simply be filled with salty water. Coal-bearing layers of rock may also be used for storage, especially in fractures in the coals themselves. From IPCC (2005, their Figure TS-2b).

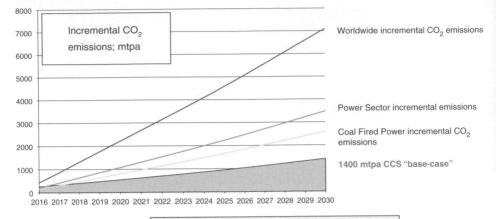

Figure 6.15 The path towards solution of a Princeton wedge with carbon capture and storage. The vertical axis shows incremental emissions of carbon dioxide in millions of tonnes per year. If half of all new coal-fired power plants were to be equipped with carbon capture and storage (CCS), that would achieve three-quarters of the target of one Princeton wedge. Slide shown by Mr Bob Edwards of BP at ING Oils Forum in London, 2 September 2008 (Kenney, 2008).

carbon dioxide this could be done, at a capital cost of the order of $20 000 million a year (Figure 6.15). This would add $0.01 – 0.05 per kilowatt–hour to the cost of electricity generated in this fashion, which compares favourably with the cost of alternatives (Table 5.2). If the value of a tonne of carbon dioxide safely stored underground were to be set at c. $50 this would be a business worth considering.

The scale of investment is enough to interest major international corporations. The estimated activity would create volumes of captured carbon dioxide equivalent to the daily volume handled by the North American natural gas market in 2008 (Figure 6.16). The greatest part of the cost (over 60%) would be in connection with capture and compression of the carbon dioxide rather than its storage (Figure 6.16).

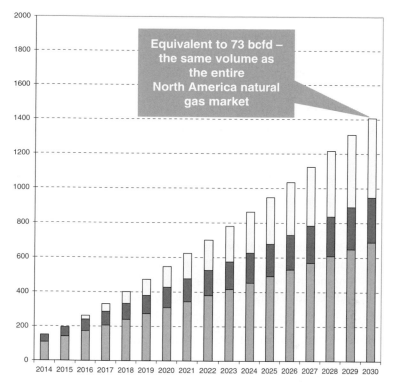

Figure 6.16 An illustrative projection of industrial volumes of carbon dioxide available for storage: lowermost section of each column is volume for USA, middle section for Europe and uppermost section for 'Other'. Note the volume of carbon dioxide of 1400 millions of tonnes per year (vertical axis) shown for the year 2030 (horizontal axis). This volume makes up the 'base case' for carbon capture and storage (CCS) featured in Figure 6.14. It is equivalent to 73 million cubic feet a day, comparable to the volume of the entire North American natural gas market. The level of annual investment required to achieve CCS at 1400 million tonnes a year by 2030 is of the order of $20 billion a year. Estimated costs of storage form only 23% of the estimated total capital expenditure; capture costs are put at 42%, compression at 25% and transport at 10%. Given the right regulatory framework, this is the scale of project to interest a major international oil company. After a BP slide shown at ING Oils Forum (Kenney, 2008).

Quick action on this development would be required in Europe, USA and China. Two Cambridge University colleagues, Martin Rees (Lord Rees, President of the Royal Society) and Mr Nick Butler (currently Senior Policy Adviser to Prime Minister

Gordon Brown, on leave from the Judge Business School) take as the title of their (2008) article in the *Financial Times*: 'Carbon capture stations must not be delayed'. They believe that our dependence on fossil fuels is likely to persist until 2050, because none of the possible alternatives will provide sufficient energy 'in time to forestall the increasing use of hydrocarbons'. They conclude: 'There seems no way to curtail the serious risk of long-term global warming unless – well before 2050 – we capture much of the carbon emitted when fossil fuels are burnt.' They illustrate the urgency by discussing a proposed new coal-fired power station in the UK:

> The decision to proceed with a new coal-fired power station at Kingsnorth should be accompanied by a decision to begin work immediately on a CCS demonstration plant in Britain. Kingsnorth's licence to operate should be limited to 10 years and extended only if CCS technology is deployed.

E.ON UK is the company seeking permission to build the power station at Kingsnorth. Mr Paul Golby, chief executive of E.ON UK, agrees with the view of Rees and Butler. Mr Mark Milner, industrial editor of *The Guardian* newspaper, reported on 18 March 2009 that Golby accepted that carbon capture would be fitted at Kingsnorth if government devised appropriate commercial arrangements:

> If they fund it, we will fit it.

Following the UK budget announcement on CCS in April 2009 (see 6.2 above) it appears that E.ON will be both required and encouraged to embrace CCS.

The essential role of the oil companies in CCS projects would be primarily to do with the later stages of compression, transport and storage, rather than the first stage of capture discussed by Golby. At least some of the technical and commercial skills required to *produce* oil and gas are comparable to those needed to inject and store carbon dioxide. A key skill is already in place. The oil industry understands porous and permeable rocks and what they contain – be those reservoirs filled with hydrocarbons, water, carbon dioxide or some combination of those fluids.

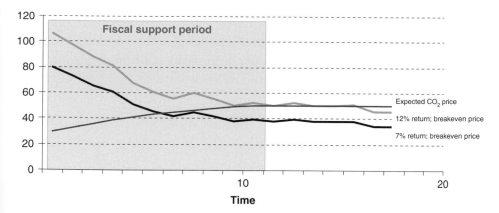

Figure 6.17 The cost of carbon dioxide capture and storage ($ per tonne, vertical axis) is currently several times greater than the reward for doing so. Given appropriate support during a transition period, learning could lead to cost reductions within ten years (time shown in years on horizontal axis). These reductions in cost, if combined with a rise in the price for *not* releasing a tonne of carbon dioxide, should give a commercial reward for the operator. After a BP slide shown at ING Oils Forum (Kenney, 2008).

So why is this business opportunity not already being seized by oil companies? The answer lies in a consideration of the time taken to reduce costs of carbon capture and storage, plotted against the time taken to develop a price for carbon dioxide that enables respectable returns to be made by those involved (Figure 6.17). Before the power industry, and indeed the cement and steel industries, start building the necessary plants, some fiscal incentives will be required from governments. These industries will need help to manage a transition across a considerable gap in costs.

That transition would lead from CCS pilot plants, incurring costs of (say) $100/tonne of carbon dioxide and losing money in a volatile and unrewarding market, to full-scale and efficient CCS plants receiving (say) $50/tonne from a mature market for carbon dioxide. That mature market can be developed only by government-imposed emissions targets and emissions caps. Yet we need to take action now: how long will it take before practical experience with carbon capture and storage by industry reduces costs enough to give a decent rate of return on investment? The answer lies in the timing of policy put in place by governments. The urgent need for such policy is the subject of the following chapter.

7

Taking it a decade at a time

7.1 GOVERNMENT REGULATION DETERMINES THE OIL INDUSTRY'S CHOICES

The actions of national governments will be crucial in providing a framework within which oil companies can implement strategies for helping to cope with climate change, particularly in relation to the carbon capture and storage discussed in the previous chapter. CCS is an essential early technology in managing the transition to a low-carbon economy by mid-century – and we need it now. Yet at the beginning of this century it would have been possible to summarise activity in this area in a single chapter of this book without running much risk of being thoroughly out of date by the time the book was published. Happily that is no longer true: a snapshot of legislation and regulation being actively developed in 2008 for carbon capture and storage is given in Table 7.1.

Whatever the upsets in the world's financial system, one hope is that the issue of energy security will help to drive an urgent government interest in several key countries in carbon capture and storage. If the generation of electricity from coal can be liberated from the incubus of the massive atmospheric pollution with carbon dioxide involved in current practices, the distribution of coal gives a new perspective on the world's future supply of low-carbon energy (Figure 7.1, Figure 7.2).

Unless there is development of low-carbon generation of electricity from coal, Europe will have to import a great deal more gas, mainly from Russia, to hit medium-term targets

Table 7.1 *A snapshot of early-2008 active legislation and regulation concerning carbon capture and storage, from data presented on 2 September 2008 by Bob Edwards of BP at ING Oils Forum (Kenney, 2008).*

North America	Canada, March	Proposal to require CCS for oil sands projects and a ban on construction of new coal plants without CCS
	USA, February	Environmental Protection Agency proposes regulations to govern injection of carbon dioxide underground
	USA, March	Legislators introduce Waxman–Markey Bill imposing a moratorium on new coal plants not equipped with CCS
	Wyoming	Approves legislation defining ownership of CCS pore space
	Oklahoma, Utah, Kentucky, Ohio	All propose or introduce or pass CCS incentives, bills or legislation
	USA	Cap-and-trade proposals in Congress provide CCS bonus allocations
Europe	EU, January	Introduces a directive specifically for CCS
	EU, January	Proposal to include CCS within the scope of the EU Emission Trading Scheme by 2012
	UK, January	Provides a regulatory framework to enable private investment in CCS projects
	EU and UK 1Q	Regulatory framework for carbon storage is being laid down
Rest of World and Global	Australia	Proposing legislation to enable underground carbon dioxide storage
	OSPAR and London Convention	Modified to include provision for carbon dioxide storage
	G8-International Energy Agency-Carbon Sequestration Leadership Forum	Recommending member governments to take comprehensive action to remove the barriers to CCS and to accelerate its development

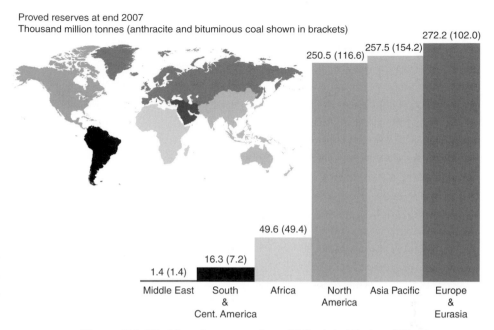

Proved reserves at end 2007
Thousand million tonnes (anthracite and bituminous coal shown in brackets)

272.2 (102.0)
257.5 (154.2)
250.5 (116.6)
49.6 (49.4)
16.3 (7.2)
1.4 (1.4)

Middle East | South & Cent. America | Africa | North America | Asia Pacific | Europe & Eurasia

Figure 7.1 World coal reserves, from *BP Statistical Review of World Energy* (BP, 2008). Note the relative lack of coal in the Middle East compared with that region's reserves of oil and natural gas, those being 61% of the world's oil and 41% of the world's natural gas (see Table 4.1 and Table 4.2; see also Figure 4.1 and Figure 4.2).

for energy supply. We cannot bank on filling the gap with electricity from renewable sources. David MacKay, Professor of Natural Philosophy in the Department of Physics at Cambridge University, is emphatic on this topic (MacKay, 2008, p. 233):

> Let's be realistic. Just like Britain, *Europe can't live on its own renewables*. So if the aim is to get off fossil fuels, Europe needs nuclear power, or solar power in other people's deserts ... or both.

Carbon capture and storage is going to be a key element in any serious attempt to cope with climate change on the scale required by the Princeton wedges (Chapter 5). Where does that leave the oil companies?

In his summary of the BP–ExxonMobil debate (Chapter 3) Mark Moody-Stuart called for increased government regulation to set the scene for coping with climate change. Here was one of the

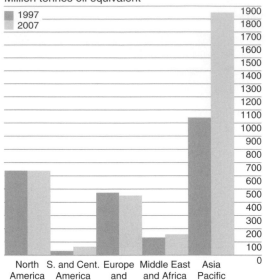

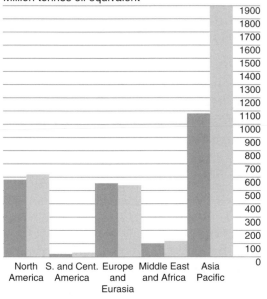

Figure 7.2 World coal production and consumption, from *BP Statistical Review of World Energy* (BP, 2008). Note the rapid growth in both production and consumption in Asia Pacific, led by China. This trend was interrupted by the global recession beginning in 2008.

world's leading oil executives asking for government regulation – a shock for traditionalists. If there is international regulation that strongly encourages the development of a low-carbon global economy, then a full range of possibilities is open to both national and independent oil companies.

At the extreme environmental end of the spectrum, with much encouragement from international regulation, the oil companies could move rapidly away from fossil fuels. If they then fail to develop areas of business comparable in scale and profit to the current oil industry they will fade away to insignificance. Others will fill the gap in supply of oil and gas for those purposes for which in the near future there are no obvious substitutes. At the environmental end of the spectrum we are considering here, the demand for oil could fall significantly, with what Dr Amory Lovins (co-founder of the entrepreneurial, non-profit Rocky Mountain Institute) sees as overall economic benefit (Lovins, 2005).

In the middle of the range, with some encouragement from national governments, oil companies could continue to explore for and produce oil and gas, while developing new low-carbon energy businesses with various degrees of urgency and commitment. The 'Masdar Initiative' by the government-owned Abu Dhabi Future Energy Company (ADFEC) is an example (Djamarani, 2008). At its core are four projects. One is building a zero-carbon, zero-waste, car-free city; another is the construction of solar-power plants. The other two projects directly involve the oil industry. A hydrogen-fired power plant with CCS is a proposed joint venture between ADFEC, BP Alternative Energy and Rio Tinto, while a nationwide CCS network costing $2–3 billion is planned for reduction of carbon emissions, combined with enhanced oil recovery.

At the other extreme, with no strong regulation of carbon dioxide emissions, the oil companies could devote all their resources to exploring the remaining frontier provinces and harvesting the remaining conventional oil and gas. Given the right price for oil, the life of the industry could be extended by the energy-intensive exploitation of oil sands and oil shale.

It is generally agreed that there can be no strong international regulation of carbon dioxide emissions without the close involvement of the government of the United States. Until the election of Barack Obama in November 2008 there appeared to be little prospect of such engagement, although hopes are now rising: 'The commitment Obama has already made to action is remarkable and gives hope to a suffering planet' is the buoyant message from the Hon. Timothy Wirth, President of the United Nations Foundation (Wirth, 2009, p. 259). And from Camilla Cavendish's column in *The Times* newspaper (Cavendish, 2009): 'In one week America has gone from playing the recalcitrant schoolboy to auditioning as the world leader on climate change.' In his inaugural speech on 20 January 2009 Obama promised that America would work with others to: 'roll back the spectre of a warming planet'. He also said that

> … each day brings further evidence that the ways we use energy strengthen our adversaries and threaten our planet.

These comments from Obama follow an intriguing straw cast into the wind in 2005 by his predecessor: the connection between energy security in the face of problems with supply of oil and gas and development of a low-carbon economy. On 30 June 2005, in an interview with *The Times* newspaper in connection with the G8 summit, this question was put to President Bush:

> On the other main G8 topic, climate change, do you believe the earth is in fact getting warmer, and if so do you believe that it is man who is making it warmer?

To which George Bush replied (this was evidently a verbatim report):

> I believe that greenhouse gases are creating a problem, a long-term problem that we got to deal with. And step one of dealing with it is to fully understand the nature of the problem so that the solutions that follow make sense.
> There is an interesting confluence now between dependency upon fossil fuels from a national economic security perspective, as well as the consequences of burning fossil fuels for greenhouse gases. And that's why it's important for our country to do two things.

One is to diversify away from fossil fuels, which we're trying to do. I think we are spending more money than any collection of nations when it comes to not only research and development of new technologies, but of the science of global warming. You know, laid out an initiative for hydrogen fuel cells. We're doing a lot of work on carbon sequestration. We hope to have a zero emissions code 5 electricity plants available for the United States as well as neighbors and friends and developing nations.

I'm a big believer that the newest generation of nuclear power ought to be a source of energy and we ought to be sharing these technologies with developing countries.

This explicit recognition in 2005 by the President of the United States that there is a problem to be faced with climate change, and that fossil fuels are part of the problem, could be read as tentative first step towards moving beyond Kyoto. Following the election of Barack Obama, decisive progress is more likely. This was anticipated by Professor Jeffrey Sachs, Director of the Earth Institute at Columbia University. He suggested (Sachs, 2007, p. 30) that the USA and other countries are moving closer to serious global negotiations to control greenhouse gas emissions, and calls for a framework that extends through to the end of the century:

The Stern Review [Stern, 2006; Professor Lord Stern is at the London School of Economics] makes clear that the costs of such control will be far lower than the costs of inaction … It is time, therefore, to aim for a sensible long-term framework in which all countries will participate. The economics are right. The U.S. Congress is set to back such a course. The White House will as well, soon after 2008 and, with some luck, even before.

What framework could be established that would enable the oil companies and others to move decisively to a low-carbon economy? John Browne and Nick Butler, both formerly at BP and now of the Cambridge Judge Business School, refer back to the ideas of John Maynard Keynes that led to the creation of the International Monetary Fund and the other institutions that fostered more than half a century of sustained growth (Browne and Butler, 2007):

> The Keynesian answer to climate change lies in the establishment
> of an international institutional structure capable of creating a
> framework within which market forces can find the most effi-
> cient solution. In short we need an International Carbon Fund.

This International Carbon Fund would set a target for emissions
reduction that would be global. Delivery would be nation by
nation according to an allocated quota. Which of the Princeton
wedges would be favoured as activities to achieve national
targets? On the economic face of it carbon capture and storage
(CCS) would not be first choice. CCS involves undeniable extra
costs, in contrast with the profit that Lovins (2005) claims can be
made from focussing on energy efficiency. But reduction in emissions from improvements in energy efficiency, and development
of renewable sources of energy, may well not be enough on their
own to achieve targets set by a Carbon Fund, especially in the
earlier years. A trading exchange designed to ensure that emissions are reduced in the most cost-effective way would allow the
market to make those choices: just what Moody-Stuart says the
oil companies need (Chapter 3).

If such regulation were to be established on this global
scale, predictions of the outcome would doubtless run a full spectrum, from the Lovins contention that it will make businesses and
consumers richer, to the apprehension of ExxonMobil (Tillerson,
2007) that we must be cautious lest prosperity be threatened.
Somewhere in the middle of the spectrum is the Stern (2006)
conclusion that the costs of controlling emissions will be much
lower than the costs of inaction.

By February 2009 Rex Tillerson was leading ExxonMobil to
acceptance of the need for intervention by government. He was
at Stanford University to celebrate six years of the Global Climate
and Energy Project, which Frank Sprow of ExxonMobil was
instrumental in establishing (Chapter 3). Tillerson told his
Stanford audience:

> One of the areas where government can provide needed stability is by
> implementing simple, transparent, and predictable policies to mitigate
> greenhouse-gas emissions … We believe that a carbon tax would be a
> more effective policy option to reduce greenhouse-gas emissions than
> alternatives such as cap-and-trade.

BP (Hayward, 2007; Reuters, 2009) is an explicit supporter of the carbon market, with, if necessary, intermediate regional arrangements on the road to global regulation (Chapter 4). Shell (Spence, 2008) also wants an international market for carbon, and like BP is prepared to build on local initiatives to achieve this. Sachs (2008, p. 40) suggests that even if it were to cost an extra 1 per cent of national income to 'head off future climate change through a more expensive energy system' we should incur that debt. That energy system might include 'carbon capture and storage technologies at coal-fired power plants or those for a large-scale solar-based electric grid'. Sachs points out that the full financial burden of this extra cost does not have to fall on us now, but could be financed through long-term government bonds to be serviced by later generations.

On the face of it, the financial crisis that began in 2008, with its extraordinary government intervention in even the most unrepentantly capitalist economies, looks like bad news for progress on environmental matters. Yet it may loosen doctrinaire economic shackles (Wirth, 2009) and encourage further development of concepts of major government financing of exceptionally long-term low-carbon projects in Europe and North America. In centrally planned economies such as that of China, there would appear to be already little restriction on such thinking.

China, which depends ever more heavily on electricity generated by burning coal, is keen to investigate technologies such as carbon capture and storage that might enable the country to maintain the use of that abundant fossil fuel, while honouring any emissions targets set by an International Carbon Fund, according to a report by science journalist Mr Jeff Tollefson (2008). A £4 million collaborative research programme of UK Research Councils with China on cleaner fossil fuels begins in 2009. On a different – but aborted – scale, in January 2008 the US Department of Energy cancelled the construction in the USA of a $1.8 billion power plant with carbon capture. The decision was not simply to cancel, but rather to use the funds to support a number of different CCS projects.

Nonetheless Tollefson (2008, p. 391) reports that this decision by the USA to cancel:

> ... baffled and angered Chinese officials and scientists ... who were partners in the project. 'This will not happen in China,' was the quote from Lu Xuedu, deputy chief of China's Ministry of Science and Technology. 'When the Chinese government says it is going to do something, it will do it, surely.'

This apparent freedom from the domestic planning controls and legal concerns that feature in other countries (Chapters 5 and 6) could lead to substantial early activity in China. In that country there could soon be commercial activity on a scale attractive to an oil company prepared to take a long view – given the right international regulatory framework.

In early 2009 carbon capture and storage is still not part of the Clean Development Mechanism (CDM) of the Kyoto Protocol. The CDM facilitates international collaboration by giving formal credit to a technologically more advanced country for establishing an approved low-carbon venture in a developing country. The reasons underlying that omission of CCS from the Clean Development Mechanism presumably include public perceptions of the difficulty of ensuring safe long-term storage of waste products. There is also a view that enabling the continued use of fossil fuels as a major source of energy is not compatible with a purist's view of a low-carbon future. Methane recovery, in contrast, can generate credits under the Clean Development Mechanism, leading to projects in China and elsewhere (Rowe, 2008).

China itself will not wish to take early action on carbon capture and storage if that entails being ruled out of billions of dollars worth of carbon credits simply because that CCS activity might be deemed to be something that 'would have happened anyway'. Meanwhile, in Europe the mess created by giving out too many credits for low-carbon ventures must be sorted out rapidly (Tickell, 2008). Despite these problems, given the necessary degree of conviction, the European experience suggests that carbon markets could be made to work under clear and firm regulation.

Governments should be impressed that after a century of resisting their regulation, the oil barons are now proposing that governments should legislate on carbon, to enable their industry to help humankind cope with climate change without the oil companies going bust in the effort. Governments must now have the guts to lead, to explain clearly to their people that together we are going to have to manage a transition to a global low-carbon economy and that this transition is bound to cost some people money in the short term, so we do not all suffer greatly in the long term. Both governments and people are now experiencing global transition, as we manage the financial crisis that became manifest in 2008. Are we thereby getting the longer perspective that we need?

7.2   LONGER PERSPECTIVES ACROSS AN OILFIELD

Oil companies are used to making decisions about rocks, fluids and pipes that involve billions of dollars and require commitment over decades rather than a few years. The life cycle of the classical oil-industry project, from frontier exploration through discovery, appraisal, development, production and abandonment, extends through much of the working life of the people involved. This contrasts with the world of politicians in a representative democracy, who have to make their decisions while paying heed to elections every few years. The oil industry is also used to reading carefully messages written in the rocks and to dealing with their implacable nature. That acceptance of a framework of activity set in stone millions of years ago comes easily to a petroleum geologist, but is naturally frustrating to those who are used to dealing mainly with supposedly malleable people rather than rigid rocks.

Some hope of a longer-range political perspective lies in the comments from George Bush concerning reliance on fossil fuels quoted in the preceding section. The issue of energy security is politically more immediate and tangible than the long-term issue of developing a low-carbon economy. Bush talks about an 'interesting confluence' between the dependence of the United States on imported oil and concerns about anthropogenic release of carbon dioxide by burning fossil fuels. President Obama's inaugural speech, quoted above, contained (no doubt unconscious) echoes of the views of his political opponent. This early-twenty-first-century interest in the relationship between energy security and the low-carbon economy may be taken as typical of the world's more powerful nations.

Whatever political fashions result from the swirling currents of world affairs, there will always be a need for the perspective that comes from an understanding of the controlling role played by Earth itself. The distribution and abundance of coal, oil and gas – and also uranium – and the consequences of the careless use of those substances, are quite unlike almost all other aspects of human affairs beyond death and taxes. Earth is not for negotiation. The distribution of rocks at and near the surface of this planet does indeed change, but only over periods of time and by inputs of energy that are not conceivably within our own reach.

The appropriate response of a rational world to the basic facts concerning climate change would be to set a long-term price for carbon and to create a fungible carbon market. In this type of market trade can take place freely through a global system of exchange, somewhat comparable to the system used to trade oil. The temptation to draw too close a parallel with the price of oil should be resisted, because that price has tended to reflect politically governed ease of access to supply rather than any global shortage of rocks containing oil. A visiting non-Earthling with a sound knowledge of the sedimentary basins containing hydrocarbons on Earth-like planets would be astonished to find us drilling in 2 or more kilometres depth of Atlantic Ocean, rather than seeking oil available at an order-of-magnitude lower cost in the deserts of Arabia.

The price of carbon has to reflect a real understanding of the danger of not controlling release of carbon dioxide. It may reasonably be claimed that we have never in human history had to create an economic structure on a global scale simply to deal with a waste product, let alone a pollutant that has no current use recognised by the public beyond the production of fizzy water. But now we have to do so, by lengthening our perspective beyond that habitually prevailing at the ballot box.

7.3   SHORTER PERSPECTIVES ACROSS A BALLOT BOX

As the world's financial crisis developed in the later part of 2008, the huge scale of government intervention in economies across the world led to a revival of interest in the works of Dr John

Maynard Keynes. His famous (1923) statement was aired: 'Long run is a misleading guide to current affairs. In the long run we are all dead.' This quotation was extended by some commentators to include the vital next sentence: 'Economists set themselves too easy, too useless a task if in tempestuous seasons they can only tell us that when the storm is past the ocean is flat again.' Far from urging that we should be passive in the face of crisis, Keynes was saying that we should not expect everything to sort itself out with the passage of time. At times governments must intervene heavily and decisively.

As I edit the final draft of this book in the spring of 2009, it is too soon to tell whether the huge intervention by governments has averted a repetition of the slump through which Keynes lived in the first half of the twentieth century. What can already be seen is that when, in 2008, the failure of poorly regulated capitalist systems became obvious to all, governments felt they could count on enough public support to intervene on a large scale. Once it becomes clear that short-term action is required to avert major long-term problems, it appears that we are indeed capable of at least considering urgent collective action across the globe. In 2008 that action included investment by governments on a scale that could make an immediately favourable short-term impression on the essential long-term development of a low-carbon economy.

So rather than seeing the early-twenty-first-century financial crisis as posing a difficulty for development of the low-carbon economy, one may be encouraged by the emphasis on agreement between governments on investment and regulation. Can this be translated by President Barack Obama and other world leaders into action on climate change? Will the fashion for government intervention help our leaders make the right decisions at the Copenhagen climate-change summit in December 2009 and beyond?

This requirement to take a view stretching over decades rather than years is frequently urged on political leaders in representative democracies. These leaders may claim that it is not easy for them to respond appropriately to such precepts, unless there is an obvious crisis that strongly influences public opinion.

Episodic heat-waves and hurricanes may be invoked by some to raise public concern on climate change, but that is a politically insubstantial path to tread. What happens if a crucial election follows a run of cool seasons? The more substantial evidence that we face a major problem with human-induced climate change is not based on instant readings of contemporary weather, and does not lend itself readily to the realities of politics at the front line – at the ballot box.

From my own experience, the voters in Edinburgh South were not sufficiently convinced by long-range views to elect an environmentally eager candidate in May 1979. Liberal Lovell came third behind Michael Ancram (later Chairman of the Conservative Party) and Gordon Brown (later Prime Minister), while the Green Party came last of five. Perhaps the thoughtful citizens of Morningside will respond more enthusiastically to any comparable offers in 2010: climate change was not in our manifestos in 1979. Now all major UK political parties are ostensibly keen to compete in generating and implementing environmentally friendly policies: in April 2009 the first official UK carbon budget was introduced. This new environmental enthusiasm in UK politicians appears to reflect significant converging interests, as discussed by Lovell (H.), Bulkeley and Owens (2009, p. 90):

> In the UK climate change and energy have converged on the policy agenda … The issue of climate change has opened up and destabilised the UK energy policy sector, but this process has been surprisingly free of conflict, despite radical policy shifts.

Is there similar convergence of policy on climate change and energy in the USA? In the summer of 2008, opinion polls in the USA suggested that issues connected with energy were more important to prospective voters in the imminent Presidential election than were matters to do with Iraq. A concern with energy in the USA at that time may have been largely to do with the price of oil and energy security: a later Pew Research poll reported in *The Times* (24 January 2009) showed a fall from 56% in 2008 to 41% in 2009 in 'American voters who regard protecting the environment as a top priority'. (The Pew Research Center is a non-partisan 'fact tank' based in Washington, DC.) But, as in the UK, there is at least a longer-term prospect in the USA of a confluence of anxiety about energy security with concerns about climate change leading to significant progress in the development of a low-carbon economy.

As oil and gas get more difficult and expensive to access, so attitudes to alternatives will change. Coal is abundant worldwide compared with oil and gas. According to the 2008 *BP Statistical Review of World Energy*, there are proven coal reserves to last 130 years at current rates of production, compared with 60 years for gas and 41 years for oil. China and India are already committed to use large quantities of coal to fuel their continuing economic development. The combination of coal with carbon capture and storage can be economic and potentially profitable. One estimate, by the Pew Center, is that a total of thirty 400-megawatt carbon capture and storage plants would cost around $30 billion, and would save between $80 and $100 billion by 2030 in abatement costs (Height, 2009).

Such favourable longer-term economics for society as a whole have to be reflected in shorter-term profits for organisations handling the carbon capture and storage. Various tribes will have to work together in a fashion familiar to the oil industry: there will be a requirement to take not too narrow a view based simply on technology. We shall have to excel at the art of the possible, just as politicians are supposed to do, while following the advice of Schumacher on economics and Owens on society (Chapter 5).

That art of the possible should include the approach commended by David Jenkins in his summary of the 2003 debate on *Coping with Climate Change* (Chapter 3). Jenkins argues for incentives rather than penalties. We have to offer encouragement to people to take seriously a huge industry devoted to conservation and waste disposal that brings them no obvious personal benefit beyond the satisfaction of knowing that they are helping to avert a long-term problem.

The public will also need to have a sense of direction provided by government. The Sceptical Environmentalist himself, Bjorn Lomborg, agrees with that (Lomborg, 2008). He echoes the words of Frank Sprow (Chapter 3) in describing the Kyoto Protocol as: 'a terribly inefficient, hugely costly way to do virtually no good'. But Lomborg then goes on to say:

> Of course, we shouldn't ignore global warming. But instead of trying to cut CO_2 emissions, we should focus on dramatically increasing the funding into energy research and development. What matters is getting low-cost low-carbon technology available faster. If the price of renewable energy dropped below the cost of fossil fuels by mid-century, everyone – including China and India – would switch to the green alternatives.

Given the numbers presented in this book (see for example Table 5.2), it is apparent that Lomborg is only partly right. Governments will have to encourage conservation and direct cutting of emissions of carbon dioxide, as well as invest in developing low-carbon technology, to have any chance of meeting the Princeton targets. Lomborg believes that we should not 'panic in our policy decisions', but surely governments must have a sense of real urgency as the rate at which we are dumping carbon dioxide into the atmosphere continues to increase (*Nature*, 2008).

It is time for us to stop dwelling on the obvious shortcomings of the Kyoto Protocol and instead strive at Copenhagen in 2009 and beyond to build on the many excellent aspects of that agreement between nations with a new and improved version. That building of new agreements must include carbon capture and storage as a major element in any new protocol, with a fungible carbon market as a means of enabling the international research, development and implementation that will be crucial to success in dealing with at least one of the required eight Princeton wedges (Figure 5.5).

The issue of a price for carbon is critical if the oil industry is going to be free to help. In a speech to the Nineteenth World Petroleum Congress in Madrid, Spain, in June 2008, BP's Chief Executive Officer Tony Hayward said that:

> … we need … to create an economy-wide carbon price. In my view, the best way of creating the conditions for reducing carbon emissions is via a cap-and-trade system.

The creation of a worldwide carbon market by introduction of appropriate regulation is emphatically in the lap of politicians. Governments must now break new ground by taking an unusually long-term view, whether or not ballot boxes are involved in their decisions. As things stand, countries such as China and India are bound to feel inhibited about going it alone with ventures such as carbon capture and storage until they can be sure that they will receive the appropriate financial credit for doing so. As discussed above in relation to China, under present regulations initiatives by individual countries might be deemed to be something that would have happened anyway and therefore not eligible for the Clean Development Mechanism under the Kyoto Protocol. Billions of dollars-worth of carbon credits could be at stake.

7.4   THE CHARACTER OF AN INDUSTRY

However short-term the political atmosphere might be now, a long view will have to be taken by the oil companies as they continue to make their choices of strategy. There are informal measures of how likely this is to happen. A harbinger might be that professional organisations with direct links to the oil industry now invite contributions on matters connected with climate change at their regular meetings. At the turn of the century those of us who had expressed public concern about the role played by the industry would not expect to be warmly received on the platform alongside those discussing frontier exploration. Now we may anticipate unsolicited invitations from the likes of PETEX London and the Scottish Oil Club, if not quite yet a call to attend an Oil Barons' Ball in Midland, Texas.

Another informal measure of how seriously the oil industry is taking climate change is not simply a count of dollars to be invested, but the quality of those asked to handle such matters. As noted in Chapter 6, when BP with Ford instigated sponsorship of the Carbon Mitigation Initiative at Princeton University, they also appointed Mr Gardiner Hill of BP (see Chapter 6) to have direct involvement in the research, bringing many years of technical and managerial experience to the project. In viewing recruitment, training and deployment of staff, the industry will have to take the longest view of all. The operational core of the oil industry relies on individuals who can combine effectively technical, financial and behavioural skills. This is not an area of activity into which so-called professional managers lacking specialist expertise can be safely introduced, unless they are muscular and well coordinated, in which case they could probably assist at a low and heavily supervised level with drilling operations and geological fieldwork. Companies must recruit and develop staff of high quality if they are to access and produce the remaining oil and gas, while making the low-carbon transition required for their survival through the twenty-first century.

The traditions of the oil industry include rugged independence in frontier territory, in difficult physical conditions, trading high personal and company risk for high possible reward. It may appear to be a long way from that excitement at the frontier to dutifully pumping carbon dioxide underground and monitoring its safe storage in an old oil or gas field. One might think that such routine transport and storage was a suitable task for a utility company – and the UK National Grid has indeed announced plans to develop a £2 billion network to pipe carbon dioxide emissions from UK power stations and store them in old North Sea gas fields (*The Times* newspaper, 11 February 2009).

My former BP colleague Nick Butler believes that, to help with CCS, oil companies should second their experts to a new service business, rather than getting involved directly themselves (Chapter 4). That may become a preferred course. Yet the oil industry also has a proud tradition of dutiful supply, of hydrocarbons and their refined products, maintained through a century of war and peace. In that time the focus of the oil explorers has evolved steadily, from field-based studies of exposures of rocks and seeps of oil (Figure 6.1), to desk-based interpretation of data largely derived by remote sensing (Figure 8.4). The adaptable oil folk have over the years reluctantly, but successfully, moved from the hurly-burly of the tent to the routine of the office. They could readily handle a further transition.

We can return to the conclusion of Chapter 4. The leadership the industry needs to effect change on the scale required can come only from deep personal conviction on two fronts:

> *We face severe economic and social costs as a global community if we continue to dump carbon dioxide into the atmosphere.*
> *The oil industry can do something substantial to help solve that problem.*

7.5   CHOICE AND CONVICTION

There is a parallel in our present-day consideration of climate change with the debate on evolution in the nineteenth century. Alfred Wallace was co-discoverer with Charles Darwin of the theory of evolution. Although famously they published this greatest of all discoveries simultaneously, it is Darwin who is recognised as the principal figure. We now know that it will never be Darwin and Wallace the way it will always be Crick and Watson. What gave Darwin the edge?

At least part of the answer is geology. Darwin was part-geologist by education and practice, Wallace was not. Wallace came to understand evolution because he developed through his observations a deep understanding of life currently on this planet and realised from its variety and distribution how species must have arisen. Yet the ultimate proof lay with Darwin. The fossils he studied showed clearly that evolution had indeed taken place: it was recorded in the rocks.

The most brilliant twenty-first-century climatologists can play the role of Wallace. Playing that part, they will then be able to tell accurately what part of future climate change is attributable to us and forecast with some confidence what lies ahead. Included in that forecast will be other changes in climate, caused by forces that are nothing to do with us. These other controls of climate change have long preceded us, and will be there long after we have become extinct, by whatever means *Homo sapiens* ceases to exist. We understand these other controls of climate change because we can study the past history of climate recorded in rocks and ice.

We also understand that we have increased the concentration of carbon dioxide in Earth's atmosphere (Figure 1.1): no reasonable person denies that this has happened. From a clear message written in the rocks 55 million years ago (Figure 7.3), we understand the changes that were experienced on this planet the last time levels of carbon dioxide in the atmosphere increased in the volume and at the rate for which we are now responsible. Those two understandings combine, in a way Darwin would approve, to cut through quantities of debate and should lead us to conviction about the choice we now face.

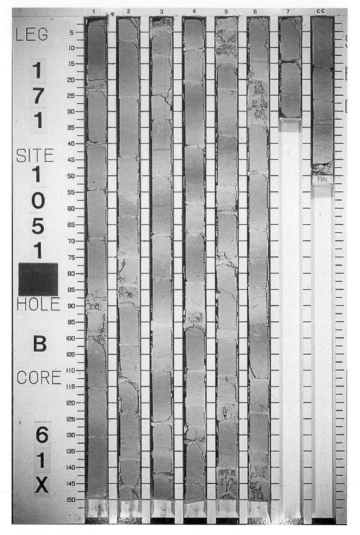

Figure 7.3 Cores of sediment (vertical scale in centimetres) from Ocean Drilling Programme Site 1051 on Blake Nose in the western North Atlantic, recovered in early 1997. It was work on the cores from Site 1051 by Norris and Röhl (1999) that brought the 55 Ma warming event onto a human timescale (Chapter 2) and thereby gave us the salutary message concerning our present activities that is a theme of this book. Photograph provided by Professor Richard Norris and Ocean Drilling Programme.

That choice cannot be ducked. It would be mighty convenient if we could pump out the remaining reserves of oil and gas and simply put the spent fossil carbon into the atmosphere, without breaching potentially dangerous limits. We cannot do this. The message from 55 million years ago is that the planet would survive: *Homo sapiens* might well not. Nor can we safely exploit the even greater reserves of useful fossil carbon in tar sands, oil shale and coal – unless we put the carbon back.

Mr Bill Spence, Vice President CO_2 of Shell, set out the position lucidly at Burlington House in London on 4 November 2008:

> In the coming 50 years, global energy demand will likely double as the population grows and increasing numbers improve their living standards. However, the reality of climate change is upon us and we will need to deliver all this energy with only half of today's carbon dioxide emissions. Twice the energy with half the carbon dioxide is a challenge for us all.

Spence went on to say that it is not sensible to spend much time arguing about which particular route we should follow to a low-carbon economy: we are going to need every reasonable approach we have to hand. Tribal behaviour won't work.

This is indeed a challenge to the established order, comparable to the greatest periods of political and social change. Successful resolution will require unprecedented cooperation between all sorts of tribes: academic, social, financial, industrial, political and national. We need the motivation that comes from reading what is written in the rocks to impel us to the action that is required. Previous ages of man have relied on rocks just as we do now. They have managed technological transitions and so shall we.

8

The proof in the puddingstone

8.1 A PERSONAL CODA

The cautionary tale that I have attempted to tell in this book is, like all stories, based to a degree on imagination. That imagination is in turn guided and constrained by some urgent messages to be read in rocks 55 million years old: messages of great significance to our grandchildren and to the state of the planet we shall pass on to them.

Plenty of things apart from charity begin at home, including this coda. In it I shall aim to summarise the theme of this book, seeking to reinforce some of the main points. I do this by tracing a circle of personal thoughts and recollection inspired by some recent archaeological and geological discoveries made in southeast England.

8.2 ROUTES AND ROCKS

A few miles from our family home in Bengeo, on the outskirts of the county town of Hertford, north of London, there is a copse on a hillside looking down on an ancient route. This Roman road became known as Ermine Street: the present-day English more prosaically call it the A10. Between the road and the copse one can walk north across a ploughed field with a left foot on 85-million-year-old (85 Ma) chalk and a right foot on 55-million-year-old (55 Ma) pebble beds. Evidence from elsewhere suggests persistence of the chalk sea in this area until around 65 million years ago, then coastal land with sea breezes – and eventually the A10. A perfect location for lines from Tennyson previously favoured by geologists: 'There, where the long street roars, hath been/ The stillness of the central sea.'

Above the field stands the copse. The farmer has not sought to bring it under the plough, and there's a good reason. The line separating chalk and pebble beds is now obscured by thicker soil and well-established trees growing between pits and mounds with relief of a few metres. Scattered on the surface between the roots lies evidence of the pebble bed below – rounded flint pebbles. Also scattered around are rarer angular fragments of rock, containing several flint pebbles tightly bound together by a cement that is as hard as the flint pebbles themselves. Any fracture-surfaces in these angular fragments cut indiscriminately across both the flint pebbles and the cement that binds them. This is the Hertfordshire Puddingstone.

Over the years the farmer has been pleased to pass to an interested local geologist the larger fragments of puddingstone turned up by the plough; Jane Tubb, who chairs the vigorous East Herts Geology Club, now has quite a collection (Figure 8.1). When the farmer's plough meets a big lump of puddingstone, his plough breaks – but puddingstone does not. Our remote ancestors, shaping it for use as querns for milling grain, dug large concretions of puddingstone (Figure 8.2) out of the largely uncemented Paleogene pebble bed in which they occur, broke them up and trimmed them in the copse – carrying or rolling the products down the hill to Ermine Street (Lovell and Tubb, 2006).

Looking down that slope from the copse, towards the many vehicles following the ancient route below, provokes some comparisons on technology. Grinding corn with puddingstone querns was more important to the survival of our Stone Age and Roman ancestors living in that area than oil is to us today. You *have* to eat: you don't really *have* to consume hydrocarbons by driving to airports and then flying to a lively Spanish resort – although that could be fun in the right sort of company. Yet even if we take the train to Spain, the farmer still needs fuel for his tractor, as do the supermarkets for their huge food trucks. Oil, unlike puddingstone, remains a mighty useful product.

Figure 8.1 Hertfordshire Puddingstone artefacts discovered by Jane Tubb. The main fragment is interpreted as a failed attempt by a disappointed Roman to make a beehive quern. Superimposed is a quarter-fragment of a previously successful attempt found nearby. Scale bars are 10 cm long. Photograph by the author, from an article in *Mercian Geologist* by Lovell and Tubb (2006).

8.3   EVIDENCE FROM OFFSHORE

Figure 8.3, a sketch I made in 1977, shows how land and sea in the area of Britain and Ireland were distributed some 80 million years ago. Hertfordshire is indeed under Tennyson's central sea, in the southeast corner of the map. Only bits of Scotland, Wales and Ireland form land; it is impossible to

Figure 8.2 'Concretion' (rounded lump of cemented rock) of Hertfordshire Puddingstone, one of an outstanding collection made by the Parkins family of High Cross, Hertfordshire, during construction of the A10 bypass that cut through their farmland. Men working on the new road, some of whom were billeted near the farm, were charged by Mrs Bessie Parkins with bringing all puddingstone recovered in operations to her for safe keeping. Scale is 50 cm. The white patch of sand seen in the centre is shown in detail in Figure 8.10. Photograph by the author, from Lovell and Tubb (2006).

recognise anything resembling present-day coastlines. That map could only be sketched with confidence once we had information from beyond those familiar present-day shores, once exploration for oil and gas beyond the coasts of Britain and Ireland really got under way in the 1960s. That exploration led to some now-legendary early successes and provoked geological liaison across tribal boundaries that persists to this day.

David Jenkins (Figure 1.5), now a hero of Chapter 1 of this book, then Chief Geologist with BP in Aberdeen, was flushed with the company's recent successful discovery of the giant Forties oilfield when, in November 1972, he came to Edinburgh to give a talk on North Sea exploration. I was his young host, then striving to be recognised as a specialist in sandstones formed in deep

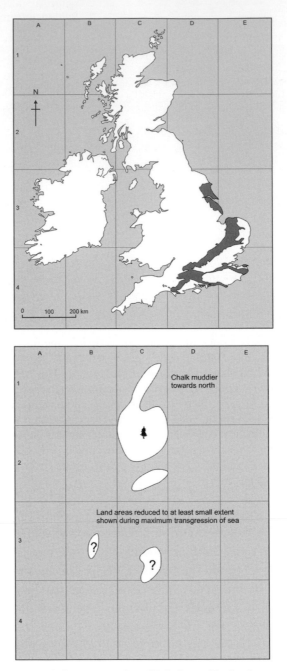

Figure 8.3 Palaeogeography of Britain and Ireland around 80 million years ago (below), with areas where Late Cretaceous rocks may be found at the surface shown (above) for reference. This sketch was made in the mid-1970s, by which time early results of drilling in the North Sea were becoming generally available. After Lovell (1977).

water, carrying out research at Edinburgh University. I had recently returned from my doctoral research with Raymond Siever at Harvard University, in an atmosphere where no one seemed snooty about either money or knowledge. Collaboration between geologists in universities and industry seemed there to be a natural part of everyday life, in a fashion then emulated in Britain only by a few pioneers such as (ex-Shell) Harold Reading at Oxford University.

A tyro is characteristically proud of his fledgling specialist abilities – in my case, to distinguish between different types of sandstone. Fresh from camping with a highly supportive field assistant in a small green tent amidst the sandstones of the Eocene Tyee Formation of the Oregon Coast Range, I was certainly not snooty about applying knowledge to practical matters. After all, the oil industry had paid for the tent. I was also more than happy to supplement a junior lecturer's salary. I listened keenly to the Chief Geologist and offered what I considered to be some quite nifty ideas over dinner with David and Evanthia Jenkins in an alarmingly expensive restaurant in Georgian Edinburgh.

Jenkins told me that there was an unresolved debate within BP concerning the deposition of the Forties reservoir sandstones: deep water or shallow water? 'The depositional environment … is still under study …' (Thomas, Walmsley and Jenkins, 1974, p. 400). Although he did not discuss it in the restaurant that evening, Jenkins already had a line of evidence indicating a relatively deep-water setting. I first became aware of this at a remarkable conference convened in Bloomsbury, London, in November 1974. At this now famous meeting, immaculately suited oil company staff mingled with dishevelled British academics. For the first time the companies were sharing previously secret information and ideas about the North Sea.

Shell had suffered a near miss with Forties, hitting only the eastern edge of the field in their own exploration acreage; but it was their man John Parker who spoke about the regional setting of this huge discovery. The rapt audience much appreciated being shown an image of the rocks lying beneath the floor of the central North Sea (Parker, 1975) (Figure 8.4). These were early days for what soon became the massive subject of 'seismic stratigraphy', the interpretation of geological history based on remote sensing

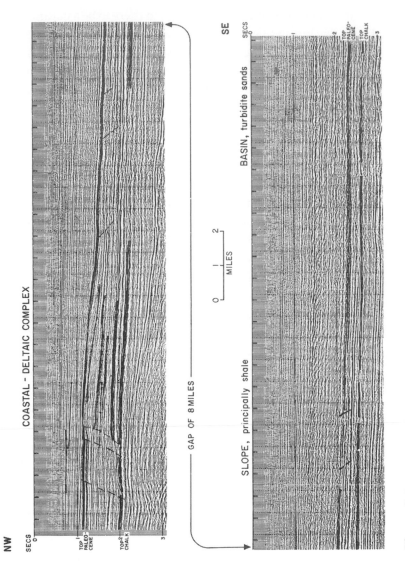

Figure 8.4

of layers of sedimentary rocks beneath Earth's surface. From my brief view of the evidence flickering across the Bloomsbury screen, I could not be sure of the exact tracing of the key lines in Parker's image, but they appeared to slope downwards from west to east, away from what could be interpreted as an ancient land mass towards the sea – and the location of the Forties reservoir sandstones. The indications were that the reservoir had been deposited in relatively deep water.

A couple of months later, in January 1975, Jenkins was back in Edinburgh. He was keen to recruit our better students from the Grant Institute of Geology into his team at Aberdeen, and into BP at large. We met in my room at the university and talked about graduate recruitment and about BP's continuing internal debate on Forties sandstones. Two days later I found myself in a store-house of cores of rock near Aberdeen's Dyce Airport, ripping the lids off wooden boxes containing sandstone cores recently cut from Forties Formation, my excitement heightening on being told that I was the first 'outsider' to look at these specimens of one of the most perfect oil reservoirs ever discovered.

Indeed the sandstones are almost too perfect to reveal their origin. They are mostly quite homogeneous, and so porous that

----

Caption for Figure 8.4 Regional cross-section of rocks below the bed of the North Sea, extending from the east coast of Scotland into the Central North Sea where Forties oilfield is located. This type of image of the large-scale arrangement of rocks below Earth's surface is created by processing remote-sensing data termed 'seismic'. Pulses of energy from airguns towed by a specially equipped ship are reflected from the sea bed and from geological features below the sea bed. The times taken for the various reflections to travel to the rocks and back are recorded by listening devices also towed by the ship. The vertical axis on the section shows this 'two-way travel time' in seconds. These data are then processed by computer and argued over by geologists and geophysicists. Forties oilfield is at the southeast end of the section, at 'Top Paleocene'. The discovery well 21/10–1 entered the top of the oil reservoir 2098 metres subsea (Walmsley, 1975). This section was first shown by John Parker of Shell at the North Sea Bloomsbury Conference in November 1974 and published the following year (Parker, 1975).

Figure 8.5 Core of Forties Formation reservoir sandstone taken from c.2755–2765 metres depth in well Forties Delta 52 on 17 May 1982, as part of a programme to assess the feasibility of enhanced oil recovery using a surfactant (a wetting agent to lower surface tension and thereby release some of the remaining oil) (see Figures 6.10 and 6.11). Photograph taken by the author with a Rollei B 35 camera which contained no electronic devices that might spark. From Lovell (2008).

they crumble readily (Figure 8.5). For me then – and now – a key part of identifying deep-water sandstones is the study of their relationship with interbedded mudstones. Not until late in the day did I open a box containing cores from the uppermost part of the main reservoir: at last – layers of sandstone grading upwards into mudstone, with a distinctive sequence of features within the sandstones and at their sharp bases. I delivered my verdict: Forties Formation was indeed deposited in relatively deep water. 'They are proximal turbidites,' I said, which stripped of jargon meant: deep-water sandstones formed relatively close to the source of supply of sand. 'So there is more sand away to the southeast – you could look for oil there,' I added. (Years later, in 1988, Enterprise Oil discovered Nelson oilfield to the southeast of Forties.) I detected a thin smile on the face of the Chief Geologist. It was almost as if Jenkins already knew the answer to the 'deep water' question.

Later that year, when the proceedings of the Bloomsbury conference appeared in print, I could see why Jenkins had smiled. 'You must have known what the seismic [Figure 8.4] showed in detail when you asked me to look at those cores,' I complained, when next we met. That thin smile again: 'I wanted to see if it was possible to reach the correct conclusion simply from the evidence within the cores.' It was. Two separate approaches had given the same answer, which in turn reduced BP's risks in planning production from Forties, and guided new exploration.

8.4   OIL MEETS ACADEMIA

That crucial final box of cores of Forties sandstone in Aberdeen changed the life of my young family. Encouraged by this happy experience of the usefulness of both knowledge and money, my wife Carol and I formed a consultancy group with my Edinburgh University colleagues Brian Price and Terry Scoffin. Then in 1981 I accepted an invitation from BP to join the oil industry full-time and moved south, into Hertfordshire Puddingstone territory. So it was that, 14 fascinating years later, in May 1995, I was visiting BP's Aberdeen office, enjoying talks on their research being given by Nicky White (Figure 2.4) and his young team from Cambridge University (Chapter 2).

They explained that Scotland had first been lifted up some hundreds of metres out of the chalk sea around 60 million years ago, as a result of the intrusion of magma that solidified to form a wedge of igneous rock a few kilometres thick at the base of the crust (Brodie and White, 1995). According to the Cambridge team, the classical geological history recorded in the glorious scenery of the Paleogene volcanoes of the Hebrides (Figure 8.6) was only part of the story. Most of the magma had been trapped deep beneath the surface, like a permanent jack beneath a car, holding Scotland above the waves: Magma Wedges.

'How fast did that magma come in beneath Scotland?' I asked White.

'About the speed of a Cambridge cyclist – to quote Dan McKenzie,' came the reply.

Figure 8.6 The roots of ancient volcanoes: Paleogene igneous rocks of the Isle of Skye, Scotland. The threatening majesty of the Cuillins is here viewed from the southeast across Loch Scavaig. Photograph by Dr Heather Lovell, 12 May 2005. From Lovell (2008).

'Did those cyclists arrive in batches?' I asked. 'That is an odd question,' replied White. 'Why?'

'Because if they didn't arrive in batches, your explanation cannot be entirely correct,' I ventured. 'We know that the North Sea sand-rich submarine fans [piles of deepwater sediments] of that age formed episodically, with muds deposited between them. So there must have been a series of upward movements of Scotland around 60 million years ago, not just one big heave.'

We already knew from BP's detailed study of the North Sea that the Forties sandstones formed one of a series of sand bodies shed from ancient Scotland. These separate sheets of sand were laid down on the floor of the ancient North Sea in a number of episodes, at intervals of a million years or more, at times when the land was high and the coast lay relatively far to the east (Stewart, 1987) (Figure 8.7). In between these times of uplift of early Scotland, the land was lower and the shore of the proto-North Sea was further to the west. The supply of sand into deeper waters was less and muds were deposited there – in time forming

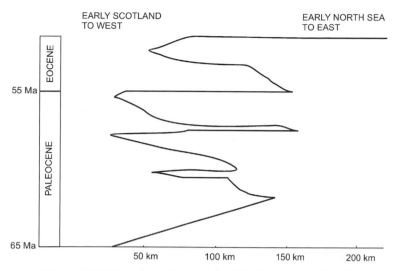

EARLY SCOTLAND
TO WEST

EARLY NORTH SEA
TO EAST

EOCENE

55 Ma

PALEOCENE

65 Ma

50 km   100 km   150 km   200 km

Figure 8.7 Advances and retreats of the Paleogene shoreline on the western flank of the ancient North Sea, reflecting episodes of uplift of early Scotland. The line follows the position of the ancient east coast of Scotland. The more detailed original diagram after which this figure was sketched was first shown at the North Sea Barbican Conference in October 1986 by Dr Ian Stewart of BP, who passed a copy to me for use in oil-industry training (Lovell, 1989). Stewart's talk at the Barbican was published the following year (Stewart, 1987).

the seals that now prevent oil trapped in the sandstones from leaking to the surface. What we in the oil gang did not know in 1995 is just *how* Scotland had been pushed up out of the chalk sea. This was what White had presented to us.

White and I published our ideas (White and Lovell, 1997) and then began to seek a fuller understanding of how these pulses of sand originated and of their fundamental cause. We came to realise that control of regional uplift by episodic intrusion of Magma Wedges at the base of Earth's crust, which we had favoured as an hypothesis in the years following our 1995 encounter, was only one particular example of a more general phenomenon. A decade later, during doctoral research funded at Cambridge by BP, our colleague Max Shaw Champion recognised evidence in his work and that of John Underhill (ex-Shell) and Fridbjorg Biskopsto at Edinburgh University, for successive uplift of the sea floor

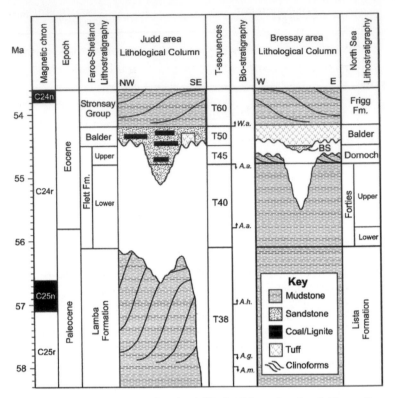

Figure 8.8 Sketch of rocks drilled offshore Scotland, Figure 2 from Rudge, Shaw Champion, White, McKenzie and Lovell (2008). The white gaps in the two columns indicate uplift of the area of Scotland at 55 million years, first (and more) to the west (Judd area) and then a little later (and less) to the east (Bressay area). Divisions (such as the 'T' numbers) are recognised by 'lithology' (type of rock), by 'biostratigraphy' (fossils) and magnetic properties. 'Clinoforms' are large-scale structures within a stack of rocks that may be detected by remote sensing ('seismic', see Figure 8.4) and used to help establish the geological history of an area. The exasperated reader will realise that, as noted in the Preface, we geologists emulate doctors and lawyers in seeking to protect our income by the use of obscure language. Reprinted with permission of Elsevier.

above the waves of the Paleogene sea in two different areas: first to the west of Scotland and then, 0.3–1.6 million years later, to the east of Scotland (Figure 8.8) (Rudge, Shaw Champion, White, McKenzie and Lovell, 2008; Shaw Champion, White, Jones and Lovell, 2008). This transient uplift took place around 55 million years ago. Peak uplift was at least 490 metres in the west, and some 300 metres in the east.

Dan McKenzie listened to Shaw Champion's explanation of this difference in timing of uplift across Scotland at one of the Bullard Laboratories' regular Friday afternoon research seminars. Shaw Champion showed how he could generate a set of numbers for the timing and extent of the uplift, and identify a probable cause – a Hot Blob in Earth's mantle, travelling from west to east deep below Scotland as a result of convection in Earth's interior. At Dan's suggestion, postdoctoral fellow John Rudge took Max's numbers and generated a quantitative model of the flow involved. Geology had guided geophysics, and in return geophysics repaid geology handsomely, with an understanding of a fundamental control of high-frequency changes in the elevation of Earth's surface. Could we now explain those numerous advances and retreats of the sea seen in the geological record for which a cause (in non-glacial times) had so long been elusive?

The Rudge–Shaw Champion Hot Blob caused transient uplift of Paleogene Scotland, leading to first retreat of the sea from the land (regression) and then its readvance (transgression) – all on an impressive scale. As uplift took place and more land emerged from the retreating Paleogene sea, rivers carried quantities of sand east to the shores of the early North Sea. Those sands were then carried further offshore to form the Forties sandstones. The volume is impressive, estimated at over 3000 cubic kilometres for the Balmoral–Fortics submarine fan of sediment alone (Reynolds, 1994): plenty of pore space in which carbon dioxide might be stored (Chapter 6). Once again, this uplift of Scotland took place during that special time in Earth history – 55 million years ago, the probable age of the Hertfordshire Puddingstone.

Coincidence? The Scottish uplift was also felt in England. With our new information we can sketch with more understanding the geography of the London and Hampshire basins of 55 million years ago (Figure 8.9). We see a recognisable outline of present-day Britain and Ireland emerging from the chalk sea, caused by a combination of Magma Wedges and Hot Blobs. And Hertfordshire lay right on that Paleogene coastline – just where you might expect

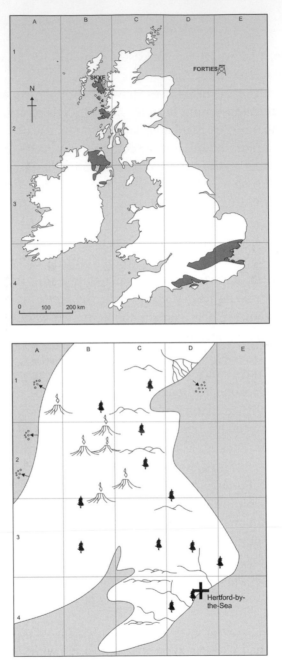

Figure 8.9 Palaeogeography of Britain and Ireland around
55 million years ago (below), showing a coastline hinting strongly
at the present-day configuration of land and sea. Areas where
Paleogene rocks may be found at the surface shown (above) for
reference. Sketch made in the mid-1970s, using data that had then
recently become available from the North Sea, including
information from Forties oilfield, the position of which is shown.
After Lovell (1977).

Figure 8.10 Hertfordshire Puddingstone – a handy pebbly beach for North Londoners 55 million years ago: close-up of the patch of uncemented fine white sand lying on the surface of the concretion shown in Figure 8.2. Coin is 25 mm across. Photograph by the author, from Lovell and Tubb (2006).

to find white sand and rounded pebbles on warm beaches (Figure 8.10).

Meanwhile, on the ocean floor away to the west, in the developing North Atlantic Ocean, an episode of dramatic global climate change was being recorded in deep-sea sediments (Figure 7.3). Thanks to the international programme of deep-sea drilling, we can now read that record 55 million years later. This brings the story onto a human timescale. As discussed in Chapter 2, we are thereby led to some uncomfortable conclusions about the effects of burning all that North Sea oil.

We are taken back to the discussion in Aberdeen in May 1995 that I had with Nicky White, which led within months to our identifying pulses of uplift and sand deposition in the region of Paleogene Scotland. As a result of episodic pulses of heat in a nearby 'hotspot' on Earth's surface that later became Iceland, pulses of sand were shed from the uplifted early Scottish land mass. These pulses of sand were linked to Hot Blobs and Magma

Wedges created by pulses of heat that made the early Iceland hotspot even hotter than usual from time to time. One major pulse of uplift occurred 55 million years ago. To the east of Scotland it created the body of sand that later became the sandstone reservoir for the 4 billion barrels of oil trapped in the Forties field. To the west of Scotland, that same pulse may have destabilised methane hydrates on the flanks of the developing North Atlantic Ocean, leading to massive release of carbon to the atmosphere and thereby triggering a major global warming event (Maclennan and Jones, 2006) (Chapter 2).

On the one hand we have a large volume of oil, a significant and famous part of one of the world's notable oil provinces. On the other we find a possible trigger of dramatic climate change, a cautionary tale from geological history. It seems to be telling us: 'Here is what happens when you release large volumes of carbon.'

At this stage my hero, Socrates, might ask: 'Is this not a natural process? What is so special about this element carbon that you make such a fuss?'

8.5   SOCRATES, CARBON AND HORSES FOR COURSES

Half a century ago, a few creative geologists imagined that the carbon-rich mudstones exposed on the shores of Dorset and Yorkshire might continue away to the north and east, beyond the present-day coast of England and Scotland. To the north and east now lies – of course – the North Sea. We can count the Kimmeridge Clay Formation as one of the most valuable rocks on the planet, in dollars as well as in stratigraphy.

Oil was generated from the carbon-rich Kimmeridge Clay Formation as it heated up during its burial beneath layer upon layer of younger rocks in the North Sea sedimentary basin. This oil made its way up from below, migrating through pores and fractures into the ample and well connected pore space in the reservoir sandstones of Forties Formation, where it was trapped under a gentle arch of rock and sealed by overlying mudstones (Figure 8.11): source, reservoir, trap and seal – and four billion barrels of oil in place. But the rapture of at least some of those who discovered, and recovered, that spectacular wealth must

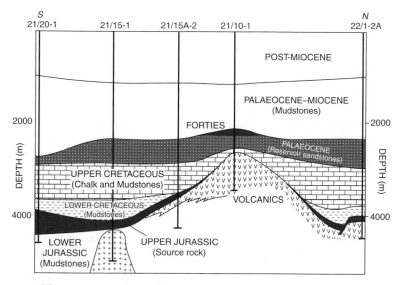

Figure 8.11 Schematic cross-section of the regional setting of Forties oilfield, indicating migration of oil from the underlying 'Upper Jurassic source rock' – the Kimmeridge Clay Formation. Note the position of discovery well 21/10–1, referred to in the caption to Figure 8.4. The sketch is taken from a BP internal report of 1985 and was made available to me at the time by Dr Tom Quigley for use in oil-industry training (Lovell, 1989).

now be tempered by the realisation that they are responsible for starting a piece of unfinished business with Earth's carbon cycle.

Although getting carbon out of the ground is expensive, the customer is happy to pay because the product is so useful. But the customer has not paid for another cost in that process, the sum of which has only recently been fully recognised. The combination of hard-won fossil carbon with atmospheric oxygen bears a heavy penalty in Earth Court, which cannot simply be discharged with payment of a single fine or a limited period of community service. We have greatly accelerated the rate of release of fossil carbon. The unfinished business is to control that rate of release to safe levels.

We have seen in Chapter 5 that Princeton University's Stephen Pacala and Robert Socolow (2004) do their oil and automobile industry patrons no special favours. Their conclusion is that keeping the concentration of atmospheric carbon dioxide at a level not far above that already reached requires the application of technology on an heroic scale. The good news is that the technology is familiar; the issue is not one of innovation, but of scale and motivation.

The numbers suggest that we do not have the luxury of choice between consuming less fossil fuel on the one hand, or carbon capture and storage on the other. Socolow and Pacala tell us that we need to do a lot of both, to have any hope of holding levels of carbon dioxide in the atmosphere at 550 parts per million by the middle of this century. As we all cope with that demand, oil geologists themselves may not feel particularly in need of redemption, despite the obloquy dished out to them by many environmentalists. But if the oil gang does feel abashed, help may lie close at hand, in their very own reservoirs. Reservoirs, like the famous sandstones containing Forties oil, that the oil industry understands so well that they can safely put the carbon back.

But is a new, giant industry – equivalent in size to that currently devoted to oil *extraction* at the rate of some 80 million barrels a day – really going to be created, to pump carbon back underground, in the form of carbon dioxide? The prospect becomes slightly less preposterous when we consider that the oil industry produces far greater volumes than just the oil itself. About three-quarters of all production from the world's oil wells is not oil at all, but brine. Include this and the total flow is over 300 million barrels per day.

Pumping 300 million barrels of compressed carbon dioxide into underground storage each day would achieve most of the Princeton target. But though the potential for a significant contribution clearly exists, 300 million barrels a day looks like an oil pipe-dream. Something much less than that volume would still make a significant contribution to the target. Why should the oil industry not seize this opportunity?

Pumping waste into long-term storage is not what we veteran frontier explorers are used to, with our techno-gambler culture of high risk and high reward (Figures 8.12 and 8.13). This would be a future service industry, with a price per tonne for all carbon safely stored. The dull psalm of duty would appear to

Figure 8.12 The thrill of the chase: Dave Pighin, a characteristically restless frontier exploration geologist, panning for gold in the mountains of British Columbia, Canada. It was a Sunday in the summer of 1980: a free day for Dave, Art Hagen and me in the midst of looking for lead, zinc and silver on behalf of Cominco. Photograph by the author, from Lovell (2008).

replace the trill of pleasure – but that is to set the technical challenges too low. The reservoir geology and engineering involved are interesting enough to quicken the blood of skilled young people. The task could be tackled properly between now and mid-century, given the motivation from a reading of the events 55 million years ago: especially what then happened to life on our planet. We need all that motivation to lend urgency to the task, if subsurface storage of carbon dioxide is to fulfil its

Figure 8.13 Meeting to review exploration operations, British Columbia, summer Sunday, 1980. Photograph by the author.

considerable potential to help us make the transition to a low-carbon economy.

The Paleocene–Eocene Thermal Maximum (PETM) at 55 Ma affected evolution to the point of defining the boundary of an Epoch. Though its significance appears reduced in comparison with the drama of the closely preceding Era-ending extinctions of dinosaurs and ammonites at 65 Ma, it was a time of major change in the evolution of mammals. There was another effect on mammals at the PETM, apart from the vicious change in climate. Thanks to uplift caused by the 55 Ma Hot Blob in the early Iceland hotspot, distant ancestors of the present-day thoroughbred racehorse were able to browse their way from one side of the nascent North Atlantic Ocean to the other over a land bridge (Hooker, 1996). So it seems appropriate that the headquarters of British racing is at Newmarket, where the gallops stretch out over the well-drained chalk downs, formed from ooze on the Cretaceous sea floor, and first lifted up to form Cambridgeshire and Suffolk as part of the development of the early Iceland hotspot.

Newmarket is less than an hour by internal combustion engine (rather than horse) from the Hertfordshire Puddingstone quarry where this coda began. The sediments that were eventually laid down on top of the puddingstone were muds – the London Clay, through which many of the capital's tunnels are cut. From the 'early hippo' bones and other fossils contained in the London Clay, we know that the climate of Bloomsbury was even steamier some 50 million years ago than when the Woolfs lived there or geologists met to exchange North Sea secrets in 1974; yet it was already considerably cooler than during the warming of the PETM, 55 million years ago.

Is there a direct link between that intense heat, and that unusual rock, the Hertfordshire Puddingstone? We cannot be sure; we cannot achieve in these Hertfordshire rocks the precision in dividing geological time developed by Nick Shackleton and his colleagues elsewhere (Chapter 1). But silica, with which the puddingstone is cemented, is more soluble at higher temperatures (Pettijohn, Potter and Siever, 1987): one may speculate that the cement that bound pebble and sand together into the plough-smashing menace of Hertfordshire farmers was formed as a result of the intense warming at the PETM. Here may be the proof in the puddingstone:

> Cos we're cooking up this planet
> Though we won't stand the heat.
> And the proof is in the puddingstone
> Right beneath our feet.
>
> The planet wasn't pleasant then, the story's in the stones.
> The oceans turned to acid – killed the fish, dissolved the bones.
> A hundred thousand years it took to clear this toxic brew;
> The climate turns quite nasty when there's too much CO_2.

In his ballad *The Puddingstone Song*, Hertfordshire singer and composer Mike Excell (2009) honours a tradition of folk music in linking apparently parochial affairs to global preoccupations. Echoes of his voice and guitar should travel beyond The Woodman pub at Chapmore End: a salutary tale from times past to be heeded in times present.

8.6 CLOSING THE CIRCLE

The circle closes. Heat from Earth's interior temporarily increased at the hotspot that marked the location of Iceland in the early North Atlantic. A pulse – a Hot Blob – in Earth's mantle convection at 55 Ma lifted Paleogene Scotland above the waves. Erosion created sand that was poured onto its flanks. That sand, carried further offshore into the precursor of today's North Sea, was finally covered and sealed by layers of mud. Later still it filled with oil from below. And there it waited until 1970, when BP came along, and drilled exploration well 21/10-1.

On proto-Scotland's west flank the same Hot Blob may have triggered the PETM, the 55 Ma warming event. Thus the same event that gave us a famous oil reservoir may also be sending us a warning about what will happen if that oil is not used wisely.

How will the Age of Oil end? All around the puddingstone copse where we began, flints lie scattered with fragments of puddingstone, unworked. Sheikh Yamani, the former oil minister of Saudi Arabia, famously said that the Stone Age didn't end because *Homo sapiens* ran out of stone. As you stoop to pick up a flint artefact (Figure 8.14) by the side of Ermine Street, with the roar of modern traffic close at hand, this observation becomes real. Like flint and puddingstone before, oil is mighty useful, and, like those rocks, much of it may remain unused. It would be great if we could with impunity produce and consume the remaining oil in our present insouciant style, but calculations suggest otherwise.

The Stone Age came to an end because our ancestors adapted to technological advance. We might suppose that we can easily do the same, that we even possess some cultural advantage over Stone Age people that assures us of emulating their success in invention and adaptation. Stephen Jay Gould (1996, p. 220) plucks that comfort from us:

Figure 8.14 Flint artefact found at the side of the Roman road now known as Ermine Street, Hertfordshire, England. Coin is 20mm across. Photograph by the author.

The Cro-Magnon people who painted the caves at Lascaux and Altamira some fifteen thousand years ago are us – and one look at the incredible richness and beauty of this work convinces us, in the most immediate and visceral way, that Picasso held no edge in mental sophistication over these ancestors with identical brains.

Rocks and their messages, like the proof in the puddingstone, may be implacable; but the man-made discipline of economics is not. Like all human culture it is a malleable product. Our brains may be unaltered in capacity since the Stone Age, but we can aspire to the creativity of the artists at Lascaux. Economics can be changed to help us to effect the transition we have to make, in response to what rocks tell us: echoes of E. F. Schumacher and his (1973) classic discussed in Chapter 5 – *Small is Beautiful: A Study of Economics as if People Mattered*.

Many might assume that we geologists must feel rather lofty and detached about climate change because we know that the planet has seen it all – and worse – before. I don't feel very lofty or detached. I am part of a large, tribal family and I have grandchildren. Our own children were young when I ran for Parliament. In the chilly church halls of Edinburgh South, I set out an earnest programme to deal with global as well as local concerns: hunger, thirst, pestilence, inequality and wickedness in high places. Were I to run for election again, I would add emphatically the warning that we can now read in those 55 million year old rocks. I reckon the undoubted evils against which I inveighed in the 1970s can only become even more pressing if, by our own hand, we create our own extreme warming event. The time in which we now live would then, sadly and justly, surely become known as the 'Anthropocene'.

We have received an important message from a warm planet. We can understand it, and we should respond – *as if people mattered.*

References

Archer, D. (2008). *The Long Thaw: How Humans Are Changing the Next 100 000 Years of Earth's Climate*. Princeton, NJ: Princeton University Press.

Archer, D., B. Buffet and V. Brovkin (2008). Ocean methane hydrates as a slow tipping point in the global carbon cycle. *Proceedings of the National Academy of Sciences*, www.pnas.org/cgi/doi/10.1073/pnas.0800885105, 1–6.

Baines, S. J. and R. H. Worden (2004a). (Eds.) *Geological Storage of Carbon Dioxide. Geological Society London Special Publications*, **233**.

Baines, S. J. and R. H. Worden (2004b). The long-term fate of CO_2 in the subsurface: natural analogues for CO_2 storage. In S. J. Baines and R. H. Worden (Eds.) *Geological Storage of Carbon Dioxide. Geological Society London Special Publications*, **233**, 59–85.

Bickle, M., A. Chadwick, H. E. Huppert, M. Hallworth and S. Lyle (2007). Modelling carbon dioxide accumulation at Sleipner: implications for underground carbon storage. *Earth and Planetary Science Letters*, **255**, 164–176.

Bierbaum, R., J. P. Holdren, M. MacCracken, R. H. Moss and P. H. Raven (2007). *Confronting Climate Change: Avoiding the Unmanageable and Managing the Unavoidable*, final report of the Scientific Expert Group on Climate Change and Sustainable Development. New York: United Nations Foundation and Sigma Xi.

BP (2008). *BP Statistical Review of World Energy*. London: British Petroleum plc.

Brodie, J. and N. J. White (1995). The link between sedimentary basin inversion and igneous underplating. In J. G. Buchan and P. G. Buchanan (Eds.) *Basin Inversion. Geological Society London Special Publications*, **88**, 31–38.

Browne, E. J. P. (1997). *Climate Change: The New Agenda*, a presentation to Stanford University, California, 19 May 1997. London: Group Media and Publications, British Petroleum plc.

Browne, J. and N. Butler (2007). We need an international carbon fund. *Financial Times*, 15 May.

Caldeira and twenty-four others (2007). Comment on 'Modern-age buildup of CO_2 and its effects on seawater acidity and salinity' by Hugo

A. Loaiciga. *Geophysical Research Letters*, **34**, L18608, doi:10.1029/2006GL027288, 2007.

Cane, M. and P. Molnar (2001). Closing of the Indonesian seaway as a precursor to East African aridification around 3–4 million years ago. *Nature*, **411**, 157–162.

Casten, T. R. and P. F. Schewe (2009). Getting the most from energy. *American Scientist*, **97**(1), 26–33.

Cavendish, C. (2009). Obama surges ahead in the race to be green. *The Times*, 6 February.

Chadwick, A., R. Arts, C. Bernstone, F. May, S. Thibeau and P. Zweigel (2008). *Best Practice for the Storage of CO_2 in Saline Aquifers*. Keyworth, Nottingham: British Geological Survey.

Cohen, A. S., A. L. Coe and D. B. Kemp (2007). The Late Palaeocene–Early Eocene and Toarcian (Early Jurassic) carbon isotope excursions: a comparison of their time scales, associated environmental changes, causes and consequences. *Journal of the Geological Society*, **164**, 1093–1108.

Cope, G. (2008). Canada – oil sands: political football. *Petroleum Review*, **62** (739), 30–32.

Cope, G. (2009). Canada – sailing stormy seas. *Petroleum Review*, **63**(744), 28, 29, 31.

Dickens, G. R. (1999). Carbon cycle: the blast in the past. *Nature*, **401**, 752–755.

Dickens, G. R., M. M. Castillo and J. C. G. Walker (1997). A blast of gas in the latest Paleocene: simulating first-order effects of massive dissociation of oceanic methane hydrate. *Geology*, **25**, 259–262.

Dickens, G. R., J. R. O'Neill, D. K. Rea and R. M. Owen (1995). Dissociation of oceanic methane hydrate as a cause of the carbon isotope excursion at the end of the Paleocene. *Paleoceanography*, **10**, 965–972.

Dixon, A. J., C. J. Beer, C. Dempsey, M. A. Maslin, J. A. Bendle, E. L. McClymont and R. D. Pancost (2009). Oceanic forcing of the Marine Isotope Stage 11 interglacial. *Nature Geoscience*, **2**, 428–433.

Djamarani, M. (2008). Middle East – clean energy: taking the initiative. *Petroleum Review*, **62**(739), 36, 37, 39.

EPICA community members (2004). Eight glacial cycles from an Antarctic ice core. *Nature*, **429**, 623–628.

Excell, M. (2009). *The Puddingstone Song*. Composed and performed by Mike Excell, www.cambridge.org/9780521145596.

Eyton, D. (2008). Energy challenges for the 21st century – an IOC perspective: the need for alignment between technology, business and policy. Speech by BP's Head of Research and Technology, Princeton University, 13 October.

Fitzgerald, S. (2005). Managing demand. *Geoscientist*, **15**(4), 20–21.

Foster, R. F. (1988). *Modern Ireland 1600–1972*. London: Allen Lane.

Fredriksen, K.-L. and T. A. Torp (2007). Breaking new ground. Project study: Statoil. In T. Nicholls (Ed.) *Fundamentals of Carbon Capture and Storage Technology*. London: Petroleum Economist, 136–143.

Gautier, C. (2008). *Oil, Water, and Climate: An Introduction*. Cambridge: Cambridge University Press.

Gerhard, L. C. (2004). Climate change: conflict of observational science, theory, and politics. *American Association of Petroleum Geologists Bulletin*, **88**, 1211–1220.

Gerhard, L. C. (2006). Climate change: conflict of observational science, theory, and politics – Reply. *American Association of Petroleum Geologists Bulletin*, **90**, 409–412.

Gerhard, L. C. and B. M. Hanson (2000). Ad hoc committee on global climate issues: annual report. *American Association Petroleum Geologists Bulletin*, **84**, 466–471.

Gibbins, J., S. Haszeldine, S. Holloway, J. Pearce, J. Oakey, S. Shackley and C. Turley (2006). Scope for future CO_2 emission reductions from electricity generation through the deployment of carbon capture and storage technologies. In J. Schellnhuber, W. Cramer, N. Nakicenovic, T. Wigley and G. Yohe (Eds.) *Avoiding Dangerous Climate Change*. Cambridge: Cambridge University Press, 379–383.

Gibson-Smith, C. (2007). We're hurting Britain, not saving the planet. *Sunday Telegraph*, 1 April.

Gilfillan, S. M. V., B. Sherwood Lollar, G. Holland, D. Blagburn, S. Stevens, M. Schoell, M. Cassidy, Z. Ding, Z. Zhou, G. Lacrampe-Couloume and C. J. Ballentine (2009). Solubility trapping in formation water as a dominant CO_2 sink in natural gas fields. *Nature*, **458**, 614–618.

Gingerich, P. D. (2006). Environment and evolution through the Paleocene–Eocene thermal maximum. *Trends in Ecology and Evolution*, **21**, 246–253.

Glaser, B., M. Parr, C. Braun and G. Kopolo (2009). Biochar is carbonate negative. *Nature Geoscience*, **2**, 2.

Goodwin, P., R. G. Williams, A. Ridgwell and M. J. Follows (2009). Climate sensitivity to the carbon cycle modulated by past and future changes in ocean chemistry. *Nature Geoscience*, **2**, 145–150.

Gore, A. (2006). *An Inconvenient Truth*. Emmaus, PA: Rodale Books.

Gould, S. J. (1996). *Life's Grandeur: The Spread of Excellence from Plato to Darwin*. London: Jonathan Cape.

Graham-Rowe, D. (2008). Four wheels good? *Nature*, **454**, 810–811.

Handley, L., P. N. Pearson, I. K. McMillan and R. D. Pancost (2008). Large terrestrial and marine carbon and hydrogen isotope excursions in a new Paleocene/Eocene boundary section from Tanzania. *Earth and Planetary Science Letters*, **275**, 17–25.

Hansen, J., M. Sato, P. Kharecha, D. Beerling, R. Berner, V. Masson-Delmotte, M. Pagani, M. Raymo, D. L. Royer and J. C. Zachos (2008). Target atmospheric CO_2: where should humanity aim? *Open Atmospheric Science Journal*, **2**, 217–231.

Harrabin, R. (2008). Germany leads 'clean coal' pilot. *BBC News at Ten* online, 22.27 GMT, 3 September.

Hays, J. D., J. Imbrie and N. J. Shackleton (1976). Variations in Earth's orbit: pacemaker of the ice ages. *Science*, **194**, 1121–1132.

Hayward, T. (2007). Delivering technologies via carbon markets. Speech to G8 Summit, Bundestag, Berlin, 4 June.

Height, M. (2009). CCS: ready by 2020? Report on second annual European CCS summit, London, December 2008. *Petroleum Review*, **63**(745), 32, 34.

Hesselbo, S. P., D. R. Grocke, H. C. Jenkyns, C. J. Bjerrum, P. Farrimond, H. S. Morgans Bell and O. R. Green (2000). Massive dissociation of gas hydrate during a Jurassic oceanic anoxic event. *Nature*, **406**, 392–395.

Hooker, J. J. (1996). Mammalian biostratigraphy across the Paleocene–Eocene boundary in the Paris, London and Belgian basins. In R. W. O'B. Knox, R. M. Corfield, and R. E. Dunay (Eds.) *Correlation of the Early Paleogene in Northwest Europe*. Geological Society London Special Publications, **111**, 205–218.

Houston, S., B. W. D. Yardley, P. G. Smalley and I. Collins (2007). Rapid fluid-rock interaction in oilfield reservoirs. *Geology*, **35**, 1143–1146.

Hunt, J. (2009). *Integrating Adaptation and Mitigation Solutions*, seminars on adaptation to climate change, Capability and Sustainability Centre, St Edmund's College, Cambridge, and Centre for Energy Studies, Judge Institute, Cambridge.

Intergovernmental Panel on Climate Change (1996). *Climate Change 1995*. Cambridge: Cambridge University Press.

Intergovernmental Panel on Climate Change (2005). *Carbon Capture and Storage*. Cambridge: Cambridge University Press.

Intergovernmental Panel on Climate Change (2007). *Climate Change 2007*. Cambridge: Cambridge University Press.

Jahn, F., M. Cook and M. Graham (2008). *Hydrocarbon Exploration and Production*, 2nd edn. Amsterdam: Elsevier.

Jenkins, D. A. L. (2001). Potential impact and effects of climate change. In L. C. Gerhard, W. E. Harrison and B. M. Hanson (Eds.) *Geological Perspectives of Global Climate Change*. American Association of Petroleum Geologists Studies in Geology, **47**, 337–359.

Johnson, J. W., J. J. Nitao and K. G. Knauss (2004). Reactive transport modelling of CO_2 storage in saline aquifers to elucidate fundamental processes, trapping mechanisms and sequestration partitioning. In S. J. Baines and R. H. Worden (Eds.) *Geological Storage of Carbon Dioxide. Geological Society London, Special Publications*, **233**, 107–128.

Kelemen, P. B. and J. Matter (2008). In situ carbonation of peridotite for CO_2 storage. *Proceedings of the National Academy Sciences of the USA*, **105**, 17 295–17 300.

Kemp, D. B., A. L. Coe, A. S. Cohen and L. Schwark (2005). Astronomical pacing of methane release in the Early Jurassic period. *Nature*, **437**, 396–399.

Kenney, J. (2008). *ING Oils Forum*, 2 September. London: ING Wholesale Banking.

Ketzer, J. M., B. Carpentier, Y. Le Gallo and P. Le Thiez (2005). Geological sequestration of CO_2 in mature hydrocarbon fields: basin and reservoir modelling of the Forties Field, North Sea. *Oil and Gas Science and Technology – Rev. IFP*, **60**, 259–273.

Keynes, J. M. (1923). *A Tract on Monetary Reform*. London: Macmillan.

Kharaka, Y. K., D. R. Cole, S. D. Hovorka, W. D. Gunter, K. G. Knauss and B. M. Freifeld (2006). Gas–water–rock interactions in Frio Formation following CO_2 injection: implications for the storage of greenhouse gases in sedimentary basins. *Geology*, **34**, 577–580.

Koonin, S. (2006). Getting serious about biofuels. *Science*, **311**, 435.

Kowalik, Z. (2004). Tide distribution and tapping into tidal energy. *Oceanologia*, **46**, 291–331.

KPMG International (2008). *Key Issues for Rising National Oil Companies*, report by Valerie Marcel. London: KPMG.

Kunzig, R. (2008). A sunshade for planet Earth. *Scientific American* (November), 46–55.

Lawson, N. (2006). *The Economics and Politics of Climate Change: An Appeal to Reason*. London: Centre for Policy Studies.

Lawson, N. (2008). *An Appeal to Reason: A Cool Look at Global Warming*. London: Duckworth.

Leggett, J. (1999). *The Carbon War: Dispatches from the End of the Oil Century*. London: Allen Lane.

Liu, J. and J. Diamond (2005). China's environment in a globalizing world. *Nature*, **435**, 1179–1186.

Lomborg, B. (2001). *The Skeptical Environmentalist*. Cambridge: Cambridge University Press.

Lomborg, B. (2008). Why cut one 3000th of a degree? It's absurd. *The Times*, 30 September.

Lourens, L. J., A. Sluijs, D. Kroon, J. C. Zachos, E. Thomas, U. Röhl, J. Bowles and I. Raffi (2005). Astronomical pacing of late Palaeocene to early Eocene global warming events. *Nature*, **435**, 1083–1087.

Lovell, (J. P.) B. (1977). *The British Isles through Geological Time: A Northward Drift*. London: George Allen & Unwin.

Lovell, (J. P.) B. (1989). Cenozoic. In K. W. Glennie (Ed.) *Introduction to the Petroleum Geology of the North Sea*. London: Blackwell Scientific Publications, 273–293.

Lovell, B. (1998). *Who Is in Charge? A Decade of Graduate Recruitment into BP: 1989–1998*. BP Archive, Warwick University.

Lovell, B. (2003). Oily war? *Geoscientist*, **13**(5), 12–14.

Lovell, B. (2006). Climate change: conflict of observational science, theory, and politics – Discussion. *American Association of Petroleum Geologists Bulletin*, **90**, 405–407.

Lovell, B. (2008). Proof in the Puddingstone: messages from a warm planet. *Geoscientist*, **18**(6,7,8), 14–17, 12–15, 12–15.

Lovell, B. and J. Tubb (2006). Ancient quarrying of rare *in situ* Palaeogene Hertfordshire Puddingstone. *Mercian Geologist*, **16**, 185–189.

Lovell, H., H. Bulkeley and S. Owens (2009). Converging agendas? Energy and climate change policies in the UK. *Environment and Planning C: Government and Policy*, **27**, 90–109, doi:10.1068/c0797j.

Lovelock, J. (2006). *The Revenge of Gaia*. London: Allen Lane.

Lovelock, J. (2009). *The Vanishing Face of Gaia*. London: Allen Lane.

Lovins, A. B. (2005). More profit with less carbon. *Scientific American* (September), 74–83.

Lu, J., M. Wilkinson, R. S. Haszeldine and A. E. Fallick (2009). Long-term performance of a mudrock seal in natural CO_2 storage. *Geology*, **37**, 35–38.

MacKay, D. J. C. (2008). *Sustainable Energy: Without the Hot Air*. Cambridge, UK: UIT. Available free online from www.withouthotair.com.

Maclennan, J. and S. M. Jones (2006). Regional uplift, gas hydrate dissociation and the origins of the Paleocene–Eocene Thermal Maximum. *Earth and Planetary Science Letters*, **245**, 65–80.

Marris, E. (2006). Black is the new green. *Nature*, **442**, 624–626.

Mather, T. (2005). *Carbon Capture and Storage (CCS)*, Postnote No. 238. London: Parliamentary Office of Science and Technology.

McManus, J. F. (2004). A great grand-daddy of ice cores. *Nature*, **429**, 611–612.

Milankovitch, M. M. (1941). Kanon der Erdbestrahlung und seine Anwendung auf das Eiszeitenproblem [Canon of insolation and ice age problems]. *Academy Royale Serbe, Belgrade Special Publication* **133**.

Milne, G. (2008). How the climate drives sea-level changes. *Astronomy and Geophysics*, **49**, 2.24–2.28.

Nature (2008). Carbon dioxide emissions rise to record levels. *News in Brief*, 2 October, 581.

Nicholls, T. (2007a). Back in the field: In Salah. Project study: In Salah. In T. Nicholls (Ed.) *Fundamentals of Carbon Capture and Storage Technology*. London: Petroleum Economist, 122–127.

Nicholls, T. (2007b). (Ed.) *Fundamentals of Carbon Capture and Storage Technology*. London: Petroleum Economist.

Nisbet, E. G., S. M. Jones, J. Maclennan, G. Eagles, J. Moed, N. Warwick, S. Bekki, P. Braesicke, J. A. Pyle and C. M. R. Fowler (2009). Kickstarting ancient warming. *Nature Geoscience*, **2**, 156–159.

Norris, R. D. and U. Röhl (1999). Carbon cycling and chronology of climate warming during the Paleocene/Eocene transition. *Nature*, **401**, 775–778.

Oreskes, N. (2004). The scientific consensus on climate change. *Science*, **306**, 1686.

Owens, S. (2004). Siting, sustainable development and social priorities. *Journal of Risk Research*, **7**, 101–114.

Pacala, S. and R. Socolow (2004). Stabilization wedges: solving the climate problem for the next 50 years with current technologies. *Science*, **305**, 968–972.

Pacala, S. and R. Socolow (2006). A plan to keep carbon in check. *Scientific American* (September), 50–56.

Pagani, M., K. Caldeira, D. Archer and J. C. Zachos (2006). An ancient carbon mystery. *Science*, **314**, 1556–1557.

Palmer, A. C., D. Keith and R. Doctor (2007). Ocean storage of carbon dioxide: pipelines, risers and seabed containment. *Proceedings, 26th International Conference on Offshore Mechanics and Arctic Engineering*, San Diego, paper OMAE 2007-29529.

Palmer, J., B. Boardman, C. Bottrill, S. Darby, M. Hinnells, G. Killip, R. Layberry and H. Lovell (2006). *Reducing the Environmental Impact of Housing*. Consultancy study for Royal Commission on Environmental

Pollution, 26th Report, The Urban Environment. Oxford: Environmental Change Institute.

Panchuk, K., A. Ridgwell and L. R. Kump (2008). Sedimentary response to Paleocene–Eocene Thermal Maximum carbon release: a model-data comparison. *Geology*, **36**, 315–318.

Parker, J. R. (1975). Lower Tertiary sand development in central North Sea. In A. W. Woodland (Ed.) *Petroleum and the Continental Shelf of North-West Europe*, vol. 1, Geology. Barking, Essex: Applied Science Publishers, 447–453.

Pearce, J., I. Czernichowski-Lauriol, S. Lombardi, S. Brune, A. Nador, J. Baker, H. Pauwels, G. Hatziyannis, S. Beaubien and E. Faber (2004). A review of natural CO_2 accumulations in Europe as analogues for geological sequestration. In S. J. Baines and R. H. Worden (Eds.) *Geological Storage of Carbon Dioxide. Geological Society London Special Publications*, **233**, 29–41.

Pettijohn, F. J., P. E. Potter and R. Siever (1987). *Sand and Sandstone*, 2nd edn. New York: Springer-Verlag.

Pielke, R. A. Jr (2006). What just ain't so: it is all too easy to underestimate the challenges posed by climate change (Book review). *Nature*, **443**, 753–754.

Poore, H. R., R. Samworth, N. J. White, S. M. Jones and I. N. McCave (2006). Neogene overflow of Northern Component Water at the Greenland–Scotland Ridge. *Geochemistry, Geophysics, Geosystems*, **7**, doi:10.1029/2005GC001085.

Press, F. and R. Siever (1978). *Earth*, 2nd edn. San Francisco: W. H. Freeman.

Raymond, L. (2000). A better path forward. In *Global Climate Change*. Dallas, TX: ExxonMobil.

Rees, M. and N. Butler (2008). Carbon capture stations must not be delayed. *Financial Times*, 14 September

Reinhardt, F. (2000). *Global Climate Change and BPAmoco*. Cambridge, MA: Harvard Business School N9–700–106.

Reuters (2009). Hayward says world needs a carbon price. 29 January.

Reynolds, T. (1994). Quantitative analysis of submarine fans in the Tertiary of the North Sea Basin. *Marine and Petroleum Geology*, **11**, 202–207.

Rochelle, C. A., I. Czernichowski-Lauriol and A. E. Milodowski (2004). The impact of chemical reactions on CO_2 storage in geological formations: a brief review. In S. J. Baines and R. H. Worden (Eds.) *Geological Storage of Carbon Dioxide. Geological Society London, Special Publications*, **233**, 87–106.

Röhl, U., T. Westerhold, T. J. Bralower and J. C. Zachos (2007). On the duration of the Paleocene–Eocene Thermal Maximum (PETM). *Geochemistry, Geophysics, Geosystems*, **8**, Q12002, doi: 10.1029/2007GC001784.

Rowe, M. (2008). Methane recovery projects booming worldwide. *Petroleum Review*, **62** (741), 42–43.

Royal Society (2006). *The Long-Term Management of Radioactive Waste: The Work of the Committee on Radioactive Waste Management (CoRWM)*, Policy document 01/06. London: The Royal Society.

Royal Society (2007). *Climate Change Controversies: A Simple Guide*. London: The Royal Society.

Rudge, J. F., M. E. Shaw Champion, N. White, D. Mckenzie and B. Lovell (2008). A plume model of transient diachronous uplift at the Earth's surface. *Earth and Planetary Science Letters*, **267**, 146–160, doi: 10.10.16/j.epsl.2007.11.040.

Sachs, J. D. (2007). Moving beyond Kyoto. *Scientific American* (February), 30.

Sachs, J. D. (2008). Looking after the future. *Scientific American* (November), 40.

Saeverud, I. A. and J. B. Skjaerseth (2007). Oil companies and climate change: inconsistencies between strategy formulation and implementation? *Global Environmental Politics*, **7**, 42–62.

Schiermeier, Q. (2006). Putting the carbon back: the hundred billion tonne challenge. *Nature*, **442**, 620–623.

Schiermeier, Q., J. Tollefson, T. Scully, A. Witze and O. Morton (2008). Energy alternatives: electricity without carbon. *Nature*, **454**, 816–823.

Schmidt, G. and D. Archer (2009). Too much of a bad thing. Nature, **458**, 1117–1118.

Schmidt, V. and D. A. McDonald (1979a). The role of secondary porosity in the course of sandstone diagenesis. In P. A. Scholle and P. R. Schluger (Eds.) *Aspects of Diagenesis. Society of Economic Paleontologists and Mineralogists Special Publications*, **26**, 175–207.

Schmidt, V. and D. A. McDonald (1979b). Texture and recognition of secondary porosity in sandstones. In P. A. Scholle and P. R. Schluger (Eds.) *Aspects of Diagenesis. Society of Economic Paleontologists and Mineralogists Special Publications*, **26**, 209–225.

Schumacher, E. F. (1973). *Small Is Beautiful: A Study of Economics as if People Mattered*. London: Abacus.

Schwab, F. (2007). Plunging into the debate on climate change. *Geotimes* (June), 60.

SCRAM (1980). *Poison in Our Hills: The First Enquiry on Atomic Waste Burial*. Edinburgh: SCRAM.

Shackley, S. and C. Gough (2006). *Carbon Capture and its Storage*. Farnham, Surrey: Ashgate.

Sharland, P. R., R. Archer, D. M. Casey, R. B. Davies, S. H. Hall, A. P. Heward, A. D. Horbury and M. D. Simmons (2001). *Arabian Plate Sequence Stratigraphy*, GeoArabia Special Publication No. 2. Bahrain: Oriental Press.

Shaw Champion, M. E., N. J. White, S. M. Jones, and J. P. B. Lovell (2008). Quantifying transient mantle convective uplift: an example from the Faroe–Shetland Basin. *Tectonics*, **27**, TC 1002, doi: 10.1029/2007TC002106.

Smith, A. G. and K. T. Pickering (2003). Oceanic gateways as a critical factor to initiate icehouse Earth. *Journal of the Geological Society*, **160**, 337–340.

Socolow, R. (2004). Can we bury global warming? *Scientific American* (July), 49–55.

Spence, B. (2008). *Climate Change and Shell's Response, Shell London Lecture Series, Burlington House*, 4 November. London: Geological Society.

Stenhouse, M. J. and D. Savage (2004). Monitoring experience associated with nuclear waste disposal and its application to CO_2 sequestration projects. In S.J. Baines and R.H. Worden (Eds.) *Geological Storage of Carbon Dioxide. Geological Society London Special Publications*, **233**, 1–6.

Stephens, J.C. and B. Van Der Zwaan (2005). The case for carbon capture and storage. *Issues in Science and Technology (Fall)*,

Stern, N. (2006). *Stern Review: The Economics of Climate Change*. Cambridge: Cambridge University Press.

Stewart, I.J. (1987). A revised stratigraphic interpretation of the early Palaeogene of the Central North Sea. In J. Brooks and K.W. Glennie (Eds.) *Petroleum Geology of North-West Europe, Proceedings of the 3rd Conference*, 557–576.

Svensen, H., S. Planke, A. Malthe-Sorenssen, B. Jamtvelt, R. Myldebust, T.R. Eldern and S.S. Rey (2004). Release of methane from a volcanic basin as a mechanism for initial Eocene global warming. *Nature*, **429**, 542–545.

Thomas, A.N., P.J. Walmsley and D.A.L. Jenkins (1974). Forties Field, North Sea. *American Association of Petroleum Geologists Bulletin*, **58**, 396–406.

Thomas, E. and N.J. Shackleton (1996). The Paleocene–Eocene benthic foraminiferal extinction and stable isotope anomalies. In R.W. O'B. Knox, R.M. Corfield, and R.E. Dunay (Eds.) *Correlation of the Early Paleogene in Northwest Europe. Geological Society London Special Publications*, **111**, 401–441.

Tickell, O. (2008). *Kyoto2: How to Manage the Global Greenhouse*. London: Zed Books.

Tillerson, R. (2007). *The State of the Energy Industry: Strengths, Realities and Solutions*. Opening address to CERA Week, Houston, Texas, 13 February.

Tollefson, J. (2008). Stoking the fire. *Nature*, **454**, 388–392.

Tripati, A. and H. Elderfield (2006). Deep-sea temperature and circulation changes at the Paleocene–Eocene thermal maximum. *Science*, **308**, 1894–1898.

Veil, J.A., M.G. Puder, D. Elcock and R.J. Redwell, Jr (2004). *A White Paper describing Produced Water from Production of Crude Oil, Natural Gas, and Coal Bed Methane*, prepared by Argonne National Laboratory for US Dept of Energy, National Energy Technology Laboratory, under contract W-31-109-Eng-38.

Walmsley, P.J. (1975). The Forties Field. In A.W. Woodland (Ed.) *Petroleum and the Continental Shelf of North-West Europe*, vol. 1, *Geology*. Barking, Essex: Applied Science Publishers, 477–485.

White, N. and B. Lovell (1997). Measuring the pulse of a plume with the sedimentary record. *Nature*, **387**, 888–891.

Wirth, T.E. (2009). Commentary: the United States must lead the way to a new climate deal at Copenhagen. *Nature*, **457**, 258–259.

Wright, J.D. and K.G. Miller (1996). Control of North Atlantic deep water circulation by the Greenland–Scotland ridge. *Paleoceanography*, **11**, 157–170.

Yergin, D. (1991). *The Prize: The Epic Quest for Oil, Money, and Power*. London: Simon & Schuster.

Index

29 99